Reviews of Environmental Contamination and Toxicology

VOLUME 256

More information about this series at https://link.springer.com/bookseries/398

Reviews of Environmental Contamination and Toxicology Volume 256

Editor
Pim de Voogt

Editorial Board
María Fernanda Cavieres, Valparaiso, Chile
James B. Knaak, Fort Myers, FL, USA
Annemarie P. van Wezel, Amsterdam, The Netherlands
Ronald S. Tjeerdema, Davis, CA, USA
Marco Vighi, Alcalà de Henares (Madrid), Spain

Founding Editor
Francis A. Gunther

Volume 256

Coordinating Board of Editors

PROF. DR. PIM DE VOOGT, *Editor*
Reviews of Environmental Contamination and Toxicology

University of Amsterdam
Amsterdam, The Netherlands
E-mail: w.p.devoogt@uva.nl

DR. ERIN R. BENNETT, *Editor*
Bulletin of Environmental Contamination and Toxicology

Great Lakes Institute for Environmental Research
University of Windsor
Windsor, ON, Canada
E-mail: ebennett@uwindsor.ca

DR. PETER S. ROSS, *Editor*
Archives of Environmental Contamination and Toxicology

Vancouver Aquarium Marine Science Center
Vancouver, BC, Canada
E-mail: peter.ross@vanaqua.org

ISSN 0179-5953   ISSN 2197-6554 (electronic)
Reviews of Environmental Contamination and Toxicology
ISBN 978-3-030-88139-9   ISBN 978-3-030-88140-5 (eBook)
https://doi.org/10.1007/978-3-030-88140-5

© The Editor(s) (if applicable) and The Author(s), under exclusive license to Springer Nature Switzerland AG 2021

This work is subject to copyright. All rights are solely and exclusively licensed by the Publisher, whether the whole or part of the material is concerned, specifically the rights of translation, reprinting, reuse of illustrations, recitation, broadcasting, reproduction on microfilms or in any other physical way, and transmission or information storage and retrieval, electronic adaptation, computer software, or by similar or dissimilar methodology now known or hereafter developed.

The use of general descriptive names, registered names, trademarks, service marks, etc. in this publication does not imply, even in the absence of a specific statement, that such names are exempt from the relevant protective laws and regulations and therefore free for general use.

The publisher, the authors and the editors are safe to assume that the advice and information in this book are believed to be true and accurate at the date of publication. Neither the publisher nor the authors or the editors give a warranty, expressed or implied, with respect to the material contained herein or for any errors or omissions that may have been made. The publisher remains neutral with regard to jurisdictional claims in published maps and institutional affiliations.

This Springer imprint is published by the registered company Springer Nature Switzerland AG
The registered company address is: Gewerbestrasse 11, 6330 Cham, Switzerland

Foreword

International concern in scientific, industrial, and governmental communities over traces of xenobiotics in foods and in both abiotic and biotic environments has justified the present triumvirate of specialized publications in this field: comprehensive reviews, rapidly published research papers and progress reports, and archival documentations These three international publications are integrated and scheduled to provide the coherency essential for nonduplicative and current progress in a field as dynamic and complex as environmental contamination and toxicology. This series is reserved exclusively for the diversified literature on "toxic" chemicals in our food, our feeds, our homes, recreational and working surroundings, our domestic animals, our wildlife, and ourselves. Tremendous efforts worldwide have been mobilized to evaluate the nature, presence, magnitude, fate, and toxicology of the chemicals loosed upon the Earth. Among the sequelae of this broad new emphasis is an undeniable need for an articulated set of authoritative publications, where one can find the latest important world literature produced by these emerging areas of science together with documentation of pertinent ancillary legislation.

Research directors and legislative or administrative advisers do not have the time to scan the escalating number of technical publications that may contain articles important to current responsibility. Rather, these individuals need the background provided by detailed reviews and the assurance that the latest information is made available to them, all with minimal literature searching. Similarly, the scientist assigned or attracted to a new problem is required to glean all literature pertinent to the task, to publish new developments or important new experimental details quickly, to inform others of findings that might alter their own efforts, and eventually to publish all his/her supporting data and conclusions for archival purposes.

In the fields of environmental contamination and toxicology, the sum of these concerns and responsibilities is decisively addressed by the uniform, encompassing, and timely publication format of the Springer triumvirate:

Reviews of Environmental Contamination and Toxicology [Vol. 1 through 97 (1962–1986) as Residue Reviews] for detailed review articles concerned with any aspects of chemical contaminants, including pesticides, in the total environment with toxicological considerations and consequences.

Bulletin of Environmental Contamination and Toxicology (Vol. 1 in 1966) for rapid publication of short reports of significant advances and discoveries in the fields of air, soil, water, and food contamination and pollution as well as methodology and other disciplines concerned with the introduction, presence, and effects of toxicants in the total environment.

Archives of Environmental Contamination and Toxicology (Vol. 1 in 1973) for important complete articles emphasizing and describing original experimental or theoretical research work pertaining to the scientific aspects of chemical contaminants in the environment.

The individual editors of these three publications comprise the joint Coordinating Board of Editors with referral within the board of manuscripts submitted to one publication but deemed by major emphasis or length more suitable for one of the others.

Coordinating Board of Editors

Preface

The role of *Reviews* is to publish detailed scientific review articles on all aspects of environmental contamination and associated (eco)toxicological consequences. Such articles facilitate the often complex task of accessing and interpreting cogent scientific data within the confines of one or more closely related research fields.

In the 50+ years since *Reviews of Environmental Contamination and Toxicology* (formerly *Residue Reviews*) was first published, the number, scope, and complexity of environmental pollution incidents have grown unabated. During this entire period, the emphasis has been on publishing articles that address the presence and toxicity of environmental contaminants. New research is published each year on a myriad of environmental pollution issues facing people worldwide. This fact, and the routine discovery and reporting of emerging contaminants and new environmental contamination cases, creates an increasingly important function for *Reviews*. The staggering volume of scientific literature demands remedy by which data can be synthesized and made available to readers in an abridged form. *Reviews* addresses this need and provides detailed reviews worldwide to key scientists and science or policy administrators, whether employed by government, universities, nongovernmental organizations, or the private sector.

There is a panoply of environmental issues and concerns on which many scientists have focused their research in past years. The scope of this list is quite broad, encompassing environmental events globally that affect marine and terrestrial ecosystems; biotic and abiotic environments; impacts on plants, humans, and wildlife; and pollutants, both chemical and radioactive; as well as the ravages of environmental disease in virtually all environmental media (soil, water, air). New or enhanced safety and environmental concerns have emerged in the last decade to be added to incidents covered by the media, studied by scientists, and addressed by governmental and private institutions. Among these are events so striking that they are creating a paradigm shift. Two in particular are at the center of ever increasing media as well as scientific attention: bioterrorism and global warming. Unfortunately, these very worrisome issues are now superimposed on the already extensive list of ongoing environmental challenges.

The ultimate role of publishing scientific environmental research is to enhance understanding of the environment in ways that allow the public to be better informed or, in other words, to enable the public to have access to sufficient information. Because the public gets most of its information on science and technology from internet, TV news, and reports, the role for scientists as interpreters and brokers of scientific information to the public will grow rather than diminish. Environmentalism is an important global political force, resulting in the emergence of multinational consortia to control pollution and the evolution of the environmental ethic. Will the new politics of the twenty-first century involve a consortium of technologists and environmentalists, or a progressive confrontation? These matters are of genuine concern to governmental agencies and legislative bodies around the world.

For those who make the decisions about how our planet is managed, there is an ongoing need for continual surveillance and intelligent controls to avoid endangering the environment, public health, and wildlife. Ensuring safety-in-use of the many chemicals involved in our highly industrialized culture is a dynamic challenge, because the old, established materials are continually being displaced by newly developed molecules more acceptable to federal and state regulatory agencies, public health officials, and environmentalists. New legislation that will deal in an appropriate manner with this challenge is currently in the making or has been implemented recently, such as the REACH legislation in Europe. These regulations demand scientifically sound and documented dossiers on new chemicals.

Reviews publishes synoptic articles designed to treat the presence, fate, and, if possible, the safety of xenobiotics in any segment of the environment. These reviews can be either general or specific, but properly lie in the domains of analytical chemistry and its methodology, biochemistry, human and animal medicine, legislation, pharmacology, physiology, (eco)toxicology, and regulation. Certain affairs in food technology concerned specifically with pesticide and other food-additive problems may also be appropriate.

Because manuscripts are published in the order in which they are received in final form, it may seem that some important aspects have been neglected at times. However, these apparent omissions are recognized, and pertinent manuscripts are likely in preparation or planned. The field is so very large and the interests in it are so varied that the editor and the editorial board earnestly solicit authors and suggestions of underrepresented topics to make this international book series yet more useful and worthwhile.

Justification for the preparation of any review for this book series is that it deals with some aspect of the many real problems arising from the presence of anthropogenic chemicals in our surroundings. Thus, manuscripts may encompass case studies from any country. Additionally, chemical contamination in any manner of air, water, soil, or plant or animal life is within these objectives and their scope.

Manuscripts are often contributed by invitation. However, nominations for new topics or topics in areas that are rapidly advancing are welcome. Preliminary communication with the Editor-in-Chief is recommended before volunteered review manuscripts are submitted. *Reviews* is registered in WebofScience™. Inclusion in the Science Citation Index serves to encourage scientists in academia to contribute to the series. The impact factor in recent years has increased from 2.5 in 2009 to 7.0 in 2017. The Editor-in-Chief and the Editorial Board strive for a further increase of the journal impact factor by actively inviting authors to submit manuscripts.

Amsterdam, The Netherlands Pim de Voogt
February 2020

Contents

Metalliferous Mining Pollution and Its Impact on Terrestrial and Semi-terrestrial Vertebrates: A Review 1
Esperanza Gil-Jiménez, Manuela de Lucas, and Miguel Ferrer

Fluorotelomer Alcohols' Toxicology Correlates with Oxidative Stress and Metabolism 71
Yujuan Yang, Kuiyu Meng, Min Chen, Shuyu Xie, and Dongmei Chen

Perchlorate Contamination: Sources, Effects, and Technologies for Remediation 103
Rosa Acevedo-Barrios and Jesus Olivero-Verbel

Sources of Antibiotic Resistant Bacteria (ARB) and Antibiotic Resistance Genes (ARGs) in the Soil: A Review of the Spreading Mechanism and Human Health Risks 121
Brim Stevy Ondon, Shengnan Li, Qixing Zhou, and Fengxiang Li

An Overview of Morpho-Physiological, Biochemical, and Molecular Responses of Sorghum Towards Heavy Metal Stress 155
Dewanshi Mishra, Smita Kumar, and Bhartendu Nath Mishra

Water and Soil Pollution: Ecological Environmental Study Methodologies Useful for Public Health Projects. A Literature Review 179
Roberto Lillini, Andrea Tittarelli, Martina Bertoldi, David Ritchie, Alexander Katalinic, Ron Pritzkuleit, Guy Launoy, Ludivine Launay, Elodie Guillaume, Tina Žagar, Carlo Modonesi, Elisabetta Meneghini, Camilla Amati, Francesca Di Salvo, Paolo Contiero, Alessandro Borgini, and Paolo Baili

List of Contributors

Rosa Acevedo-Barrios Environmental and Computational Chemistry Group, School of Pharmaceutical Sciences, University of Cartagena, Cartagena, Colombia
Grupo de Investigación en Estudios Químicos y Biológicos, Facultad de Ciencias Básicas, Universidad Tecnológica de Bolívar, Cartagena, Colombia

Camilla Amati Analytical Epidemiology and Health Impact Unit, Fondazione IRCCS "Istituto Nazionale dei Tumori", Milan, Italy

Paolo Baili Analytical Epidemiology and Health Impact Unit, Fondazione IRCCS "Istituto Nazionale dei Tumori", Milan, Italy

Martina Bertoldi Environmental Epidemiology Unit, Fondazione IRCCS "Istituto Nazionale dei Tumori", Milan, Italy

Alessandro Borgini Environmental Epidemiology Unit, Fondazione IRCCS "Istituto Nazionale dei Tumori", Milan, Italy
International Society of Doctors for the Environment (ISDE), Arezzo, Italy

Dongmei Chen National Reference Laboratory of Veterinary Drug Residues (HZAU) and MAO Key Laboratory for Detection of Veterinary Drug Residues, Wuhan, Hubei, China
MOA Laboratory for Risk Assessment of Quality and Safety of Livestock and Poultry Products, Huazhong Agricultural University, Wuhan, Hubei, China

Min Chen National Reference Laboratory of Veterinary Drug Residues (HZAU) and MAO Key Laboratory for Detection of Veterinary Drug Residues, Wuhan, Hubei, China

Paolo Contiero Environmental Epidemiology Unit, Fondazione IRCCS "Istituto Nazionale dei Tumori", Milan, Italy

Manuela de Lucas Applied Ecology Group, Department of Ethology and Biodiversity Conservation, Estación Biológica de Doñana, Consejo Superior de Investigaciones Científicas (CSIC), Seville, Spain

Francesca Di Salvo Pancreas Translational and Clinical Research Center, Ospedale IRCCS "San Raffaele", Milan, Italy

Miguel Ferrer Applied Ecology Group, Department of Ethology and Biodiversity Conservation, Estación Biológica de Doñana, Consejo Superior de Investigaciones Científicas (CSIC), Seville, Spain

Esperanza Gil-Jiménez Fundación Migres, CIMA (International Bird Migration Center), Tarifa, Spain

Elodie Guillaume Normandie Univ, UNICAEN, INSERM, ANTICIPE, Caen, France

Alexander Katalinic Institute for Cancer Epidemiology at the University Lübeck, Lübeck, Germany

Smita Kumar Department of Biochemistry, King George's Medical University, Lucknow, Uttar Pradesh, India

Ludivine Launay Normandie Univ, UNICAEN, INSERM, ANTICIPE, Caen, France
Centre François Baclesse, Caen, France

Guy Launoy Normandie Univ, UNICAEN, INSERM, ANTICIPE, Caen, France
Pôle recherche – Centre Hospitalier Universitaire, Caen, France

Fengxiang Li Key Laboratory of Pollution Processes and Environmental Criteria at Ministry of Education, Tianjin Key Laboratory of Environmental Remediation and Pollution Control, College of Environmental Science and Engineering, Nankai University, Tianjin, People's Republic of China

Shengnan Li Key Laboratory of Pollution Processes and Environmental Criteria at Ministry of Education, Tianjin Key Laboratory of Environmental Remediation and Pollution Control, College of Environmental Science and Engineering, Nankai University, Tianjin, People's Republic of China

Roberto Lillini Analytical Epidemiology and Health Impact Unit, Fondazione IRCCS "Istituto Nazionale dei Tumori", Milan, Italy

Elisabetta Meneghini Analytical Epidemiology and Health Impact Unit, Fondazione IRCCS "Istituto Nazionale dei Tumori", Milan, Italy

Kuiyu Meng National Reference Laboratory of Veterinary Drug Residues (HZAU) and MAO Key Laboratory for Detection of Veterinary Drug Residues, Wuhan, Hubei, China

Bhartendu Nath Mishra Department of Biotechnology, Institute of Engineering and Technology, Dr. A.P.J. Abdul Kalam Technical University, Lucknow, Uttar Pradesh, India

List of Contributors

Dewanshi Mishra Department of Biotechnology, Institute of Engineering and Technology, Dr. A.P.J. Abdul Kalam Technical University, Lucknow, Uttar Pradesh, India

Carlo Modonesi Cancer Registry Unit, Fondazione IRCCS "Istituto Nazionale dei Tumori", Milan, Italy
International Society of Doctors for the Environment (ISDE), Arezzo, Italy

Jesus Olivero-Verbel Environmental and Computational Chemistry Group, School of Pharmaceutical Sciences, University of Cartagena, Cartagena, Colombia

Brim Stevy Ondon Key Laboratory of Pollution Processes and Environmental Criteria at Ministry of Education, Tianjin Key Laboratory of Environmental Remediation and Pollution Control, College of Environmental Science and Engineering, Nankai University, Tianjin, People's Republic of China

Ron Pritzkuleit Institute for Cancer Epidemiology at the University Lübeck, Lübeck, Germany

David Ritchie Association Européenne des Ligues contre le Cancer, Bruxelles, Belgium

Andrea Tittarelli Cancer Registry Unit, Fondazione IRCCS "Istituto Nazionale dei Tumori", Milan, Italy

Shuyu Xie National Reference Laboratory of Veterinary Drug Residues (HZAU) and MAO Key Laboratory for Detection of Veterinary Drug Residues, Wuhan, Hubei, China

Yujuan Yang National Reference Laboratory of Veterinary Drug Residues (HZAU) and MAO Key Laboratory for Detection of Veterinary Drug Residues, Wuhan, Hubei, China

Tina Žagar Institute of Oncology Ljubljana, Ljubljana, Slovenia

Qixing Zhou Key Laboratory of Pollution Processes and Environmental Criteria at Ministry of Education, Tianjin Key Laboratory of Environmental Remediation and Pollution Control, College of Environmental Science and Engineering, Nankai University, Tianjin, People's Republic of China

Metalliferous Mining Pollution and Its Impact on Terrestrial and Semi-terrestrial Vertebrates: A Review

Esperanza Gil-Jiménez, Manuela de Lucas, and Miguel Ferrer

Contents

1 Introduction 2
2 Methodology 4
3 Results and Discussion 5
   3.1 Geographical and Temporary Distribution and Operational Status 5
   3.2 Main Metal Ore 30
   3.3 Sentinel Species 30
   3.4 Biomarkers 35
4 Conclusions 53
References 55

Abstract Metalliferous mining, a major source of metals and metalloids, has severe potential environmental impacts. However, the number of papers published in international peer-reviewed journals seems to be low regarding its effects in terrestrial wildlife. To the best of our knowledge, our review is the first on this topic. We used 186 studies published in scientific journals concerning metalliferous mining or mining spill pollution and their effects on terrestrial and semi-terrestrial vertebrates. We identified the working status of the mine complexes studied, the different biomarkers of exposure and effect used, and the studied taxa. Most studies (128) were developed in former mine sites and 46 in active mining areas. Additionally, although several mining accidents have occurred throughout the world, all papers about effects on terrestrial vertebrates from mining spillages were from Aznalcóllar

---

Supplementary Information The online version of this chapter (https://doi.org/10.1007/398_2021_65) contains supplementary material, which is available to authorized users.

E. Gil-Jiménez (✉)
Fundación Migres, CIMA (International Bird Migration Center), Tarifa, Spain
e-mail: egil@fundacionmigres.org

M. de Lucas · M. Ferrer
Applied Ecology Group, Department of Ethology and Biodiversity Conservation, Estación Biológica de Doñana, Consejo Superior de Investigaciones Científicas (CSIC), Seville, Spain
e-mail: manuela@ebd.csic.es; mferrer@ebd.csic.es

© The Author(s), under exclusive license to Springer Nature Switzerland AG 2021
P. de Voogt (ed.), *Reviews of Environmental Contamination and Toxicology Volume 256*,
Reviews of Environmental Contamination and Toxicology 256,
https://doi.org/10.1007/398_2021_65

(Spain). We also observed a lack of studies in some countries with a prominent mining industry. Despite >50% of the studies used some biomarker of effect, 42% of them only assessed exposure by measuring metal content in internal tissues or by non-invasive sampling, without considering the effect in their populations. Most studied species were birds and small mammals, with a negligible representation of reptiles and amphibians. The information gathered in this review could be helpful for future studies and protocols on the topic and it facilitates a database with valuable information on risk assessment of metalliferous mining pollution.

Keywords Biomarker · Metals · Mine site · Mining spill · Risk assessment · Sentinel species · Trace elements · Wildlife

1 Introduction

Historically, human activities have caused diverse effects on ecosystems, contaminating both their abiotic components and organisms (Nriagu and Pacyna 1988; Rattner 2009). Agriculture, industry, smelting, and metallic and non-metallic mining are among the major sources of metals in the environment (Sharma and Agrawal 2005; van Ooik et al. 2008). Particularly, mining is an important source of metals and metalloids with severe potential environmental impacts (UNEP 2000; Pereira et al. 2006; Satta et al. 2012), posing a risk to wildlife inhabiting, breeding, or feeding in the surrounding zones (Baos et al. 2006c; Berglund et al. 2011). Mining industry has cohabited with humans for thousands of years (Nocete et al. 2005), but between 1930 and 1985 the production of several elements increased between 2- and 35-fold (Nriagu and Pacyna 1988). Nowadays, it is estimated that there are over 30,000 mines around the world, of which over a third are currently being explored, developed, or mined (IRP 2020). In addition, the output of the global metal production in 2017 was about two billion tons (Reichl and Schatz 2019). Within mining industry, metalliferous mining comprehends the extraction and processing of metals and metalloids (referred to hereafter as metals or elements) by different methods. These processes have significantly increased the levels and bioavailability of naturally occurring elements in environmental compartments (Pereira et al. 2006; Sánchez-Chardi et al. 2007a). Some of them are physiologically essential elements, e.g. zinc (Zn), iron (Fe), or copper (Cu), which can cause pathological conditions when in deficiency, but in excess can have toxic effects (Damek-Poprawa and Sawicka-Kapusta 2003). Other elements, such as arsenic (As), cadmium (Cd), mercury (Hg), and lead (Pb), are naturally occurring in the environment but toxic for the organisms even at low levels (Drouhot et al. 2014; Camizuli et al. 2018). Thus, when they enter the food chain, they can accumulate and be hazardous for wildlife and even for humans inhabiting the surrounding area (Pereira et al. 2006; Sánchez-Chardi et al. 2009; Alvarenga et al. 2014).

Besides the fact that this kind of contamination persists for hundreds of years (Younger et al. 2002), dispersion of metal particles occurs not only during the lifetime of the mine, but also after being closed or abandoned. Mining waste is produced in all stages of mining and processing operations, with the amount of solid waste produced representing 90% or more of the material mined (Gutiérrez et al. 2016). In the past, after ceasing mining activity, many mine sites were completely abandoned and their mine tailings were left at open air without any treatment, posing several threats to the environment (Laurinolli and Bendell-Young 1996; Satta et al. 2012). These abandoned mine lands, also known as "derelict" or "orphan" mines, are referred to as "areas or sites of former mining activity for which no single individual, company or organization can be held responsible" (Venkateswarlu et al. 2016). The worldwide estimated number of abandoned mine lands is more than one million. Thus, derelict mines are nowadays considered one of the major contributors of severe degradation of the environment (Venkateswarlu et al. 2016) and abandoned wastes are among the worst environmental problems and hazard to ecosystems and human health (Gutiérrez et al. 2016). Regarding modern operational mines and their wastes, nowadays there exists a stricter legislation about mining pollution, remediation, and restoration of mining areas in many countries. However, mining is one of the main economic activities in several developing countries such as Colombia, Zimbabwe, Brazil, Ghana, or Zambia, where the environmental legislation is still scarce (Ikenaka et al. 2012; Deikumah et al. 2014).

Generally, organisms living in and around mining areas are continuously exposed to high levels of metals. Besides this chronic exposure, acute episodes such as accidental spills are of major concern because of the consequences they may have on wildlife (Grimalt et al. 1999; Baos et al. 2006c). Mining ore processing, lead smelting, and artisanal small-scale gold mining ranked the 2nd, 3rd, and 5th, respectively, among the worst polluting industries according to Green Cross Switzerland and Pure Earth (2016). In addition, the Agency for Toxic Substances and Disease Registry (ATSDR 2019) placed As, Pb, and Hg in the 1st, 2nd, and 3rd position in their Substance Priority List. Although this list is not a list of "most toxic" substances, it highlights their hazard. It is important to consider that, although a high number of laboratory experiments link metals with adverse health effects in animals, they do not reflect the variability of natural ecosystems, and therefore field studies are essential too (Damek-Poprawa and Sawicka-Kapusta 2004; Vanparys et al. 2008). There are many studies on how metals, resulting from several activities, accumulate and affect plants (DalCorso 2012) and animals (Rattner 2009). On the one hand, the latter author states that wildlife toxicology has been shaped by chemical use and misuse, ecological mishaps, catastrophic human poisonings, and research in the allied field of human toxicology. On the other hand, Mateo et al. (2016) stated that the motivation of researchers to work in wildlife ecotoxicology has largely related to the conservation of higher vertebrates, and by seeking to identify health risks in vertebrates that are closely related to us, potential hazards to humans are sometimes also highlighted. Either way, assessment of both ecosystem and human health are closely related. Because of that, it is necessary to highlight the use of certain wild animals as *sentinel species*, defined as an "animal system to identify potential health hazards to other animals or humans" (Basu et al. 2007). Thus, many studies developed in polluted areas, including mining sites, have used sentinel species for human and environmental health risk assessment (Custer 2011; Gómez-Ramírez et al. 2014; Mateo et al. 2016; Rodríguez-Estival et al. 2020). The multiple negative effects of exposure to metals on animal and human health are well known (Tchounwou et al. 2012; Mateo et al. 2016), and they have been assessed through different approaches. Historically pollution monitoring was based exclusively on chemical analyses of water, soil, sediments, or tissue samples. Nowadays, studies usually not only focus on the level of exposure or tissues levels, but also go one step beyond and make use of biomarkers of effect to assess the repercussions of these contaminants on the organisms (Kakkar and Jaffery 2005; Baos et al. 2006c; Mussali-Galante et al. 2013b). Thus, a *biomarker of exposure* is considered to be the detection of a contaminant itself or its metabolites, whereas a *biomarker of effect* is the alteration of a biochemical or physiological parameter associated with contaminant exposure (Chaousis et al. 2018). But this definition of biomarker of effect should be extended not only to the individual level, but also to effects that can be deleterious at population or ecosystem levels, such as reproductive parameters (Berglund et al. 2010; Mussali-Galante et al. 2013b).

The purpose of the present review is to analyze and synthesize the existing literature on metal pollution produced specifically by metalliferous and non-radioactive mining activities or problems derived from them, such as spillages, which affect semi-terrestrial and terrestrial vertebrates. Thus, we have focused on the temporal and worldwide distribution of these mining studies and their working status, on the different effects found according to different biomarkers and on an overview of studied taxa. There are several scientific papers about aquatic ecosystems and/or plants, but studies focused on mine products and terrestrial or semi-terrestrial animals are not as large as expected according to the high number of mines and the riskiness of this pollution on terrestrial ecosystems. To the best of our knowledge, the present study is the first review regarding effects of metalliferous mining activity on semi-terrestrial and terrestrial animals. The compilation of all scientific studies available on this topic is important to highlight the gaps and unevenness that exist in these studies. But also to establish a starting point on what is necessary to be done in the future, both scientifically and in the application of this scientific knowledge in mining industry, including regulation, policy, and decision making.

2 Methodology

We searched in Web of Science and Google Scholar for articles published or available online until the end of 2019. We used different keywords and combination of terms, such as "mining," "mine," "heavy metal," "trace element," "wildlife," "animals," "contamination," or "mining spill," and then we also checked for subsequent references as well as authors bibliographies. From our preliminary search of published literature, we identified all relevant peer-reviewed papers reporting metalliferous mining or mining spill contamination in terrestrial and semi-terrestrial vertebrates. We included waterfowl as semi-terrestrial vertebrates because they are also related to terrestrial ecosystem and are usually present at mining sites. We excluded those studies focused exclusively on metallurgical processing (e.g., only reporting smelter pollution without mineral extraction); with other pollution sources (e.g., coal mines or marble and stone quarries) or no clear specification of the source; with different pollutants like cyanide; or those studying plants or aquatic species. Our final set for assessment included 186 studies that met the criteria specified. Then, we organized the information identifying the specific source of contamination (closed or abandoned mine, active mine, spillage), the main metal(s) extracted, the year of publication, the studied species and the biomarkers utilized (exposure or effects) (Table 1). Also, when a study assessed metal pollution from different sources, we focus only on our source of interest. Some of the articles studied more than one species or included data from mines before and after being closed. All data and their references are included in Supplemental Data (Annex 1).

3 Results and Discussion

3.1 Geographical and Temporary Distribution and Operational Status

Studies were classified according to the operational status of the mine at the time when research was performed, classifying them in three main categories: closed or abandoned mines (both terms were used interchangeably in the present review), operational mines, and mining spillages. These studies were distributed over 24 countries, and they were not equally distributed worldwide for any of the groups (Figs. 1 and 2).

The temporal trend of scientific research on pollution by metalliferous mining in terrestrial and semi-terrestrial vertebrates (Fig. 3) showed an increment in the number of papers per year in the last 20 years. Particularly, between 1999 and 2009, the increase in the number of publications related to mining spillages, due to the Aznalcóllar accident (later discussed) is noteworthy. The general trend is in line with that found in scientific research on the use of bioindicators (Burger 2006) and on metal and metalloid pollution studies on wildlife in Latin America (Di Marzio et al. 2019).

Regarding abandoned or closed mines, more than half of studies about metalliferous mining pollution were related with them. 111 of 186 (60%) studies assessed pollution exclusively from former mines, 16 studies were both at a former mine and an active mine (most of them in big mining complexes), and one was developed both in an old mine and a spillage site. In total, 128 studies included an old mine site. Ten of these studies were carried out in closed mines which had undergone partially or

Table 1 Species used in the studies

Species	N	Metal ore	Biomarker	Country (continent)	Mine status	Reference
AVES						
Accipitriformes (9 species)						
Black kite (*Milvus migrans*)	5	Pb/Zn/Cu	EF	Spain (E)	Mining spill	Baos et al. (2006b)
		Pb/Zn/Cu	EF	Spain (E)	Mining spill	Baos et al. (2006c)
		Pb/Zn/Cu	EXP	Spain (E)	Mining spill	Benito et al. (1999)
		Pb/Zn/Cu	EF	Spain (E)	Mining spill	Pastor et al. (2004)
		Pb/Zn/Cu	EXP	Spain (E)	Mining spill	Rodríguez Álvarez et al. (2013)
Booted eagle (*Hieraaetus pennatus*)	1	Pb/Zn/Cu	EF	Spain (E)	Mining spill	Gil-Jiménez et al. (2017)
Marsh harrier (*Circus aeruginosus*)	1	Pb/Zn/Cu	EXP	Spain (E)	Mining spill	Rodríguez Álvarez et al. (2013)
Northern Harrier (*Circus cyaneus*)	1	Pb/Zn (Ag/Au)	EF	USA (NA)	Old	Henny et al. (1994)
Osprey (*Pandion haliaetus*)	2	Pb/Zn (Ag/Au)	EF	USA (NA)	Old	Henny et al. (1991)
		Ag/Au	EXP	USA (NA)	Old	Langner et al. (2012)
Red-tailed Hawk (*Buteo jamaicensis*)	1	Pb/Zn (Ag/Au)	EF	USA (NA)	Old	Henny et al. (1994)
Sparrowhawks (*Accipiter nisus*)	1	Cu/As	EXP	United Kingdom (E)	Old	Erry et al. (1999a)
Wild turkey vulture (*Cathartes aura*)	1	Cu	EXP	Chile (SA)	Old	Valladares et al. (2013)
Bald eagle (*Haliaeetus leucocephalus*)	1	Hg	EF	Canada (NA)	Old	Weech et al. (2006)
Anseriformes (17 species)						
Canada goose (*Branta canadensis*) [And 14 other species in Sileo et al. (2001)]	6	Pb/Zn (Ag/Au)	EXP	USA (NA)	Old	Beyer et al. (1998)
		Pb/Zn	EF	USA (NA)	Old	van der Merwe et al. (2011)
		Pb/Zn (Ag/Au)	EF	USA (NA)	Old	Blus et al. (1995)
		Pb/Zn (Ag/Au)	EF	USA (NA)	Old	Sileo et al. (2001)
		Pb/Zn	EF	USA (NA)	Old	Sileo et al. (2003)
		Pb/Zn	EF	USA (NA)	Old	Beyer et al. (2004)
Common goldeneye (*Bucephala clangula*)	1	Pb/Zn (Ag/Au)	EF	USA (NA)	Old	Blus et al. (1995)
Common pintail (*Anas acuta*)	1	Pb/Zn	EF	USA (NA)	Old	Beyer et al. (2004)

Gadwall (*Anas strepera*)	2	Pb/Zn/Cu	EXP	Spain (E)	Mining spill	Benito et al. (1999)
		Pb/Zn/Cu	EXP	Spain (E)	Mining spill	Taggart et al. (2006)
Green-winged teal (*Anas crecca*)	1	Pb/Zn	EF	USA (NA)	Old	Beyer et al. (2004)
Greylag goose (*Anser anser*)	2	Pb/Zn/Cu	EXP	Spain (E)	Mining spill	Mateo et al. (2006)
		Pb/Zn/Cu	EF	Spain (E)	Mining spill	Martinez-Haro et al. (2013)
Lesser scaup (*Aythya affinis*)	1	Pb/Zn	EF	USA (NA)	Old	Beyer et al. (2004)
Mallard (*Anas platyrhynchos*)	8	Pb/Zn (Ag/Au)	EF	USA (NA)	Old	Blus et al. (1995)
		Pb/Zn/Cu	EXP	Spain (E)	Mining spill	Rodríguez Álvarez et al. (2013)
		Pb/Zn/Cu	EXP	Spain (E)	Mining spill	Benito et al. (1999)
		Pb/Zn	EF	USA (NA)	Old	Beyer et al. (2004)
		Pb/Zn (Ag/Au)	EXP	USA (NA)	Old	Beyer et al. (1998)
		Pb/Zn (Ag/Au)	EF	USA (NA)	Old	Henny et al. (2000)
		Pb/Zn	EF	USA (NA)	Old	Sileo et al. (2003)
		Pb/Zn/Cu	EXP	Spain (E)	Mining spill	Taggart et al. (2006)
Common Pochard (*Aythya ferina*)	2	Pb/Zn/Cu	EXP	Spain (E)	Mining spill	Benito et al. (1999)
		Pb/Zn/Cu	EXP	Spain (E)	Mining spill	Taggart et al. (2006)
Red-crested pochard (*Netta rufina*)	1	Pb/Zn/Cu	EXP	Spain (E)	Mining spill	Taggart et al. (2006)
Ring-necked duck (*Aythya collaris*)	1	Pb/Zn	EF	USA (NA)	Old	Beyer et al. (2004)
Shaoxing duck (reared under natural conditions)	1	Hg	EF	China (AS)	Old	Ji et al. (2006)
Shoveler (*Anas clypeata*)	1	Pb/Zn/Cu	EXP	Spain (E)	Mining spill	Rodríguez Álvarez et al. (2013)
Trumpeter swan (*Cygnus buccinator*)	1	Pb/Zn	EF	USA (NA)	Old	Carpenter et al. (2004)
Tundra swans (*Cygnus columbianus*)	5	Pb/Zn/Ag	EXP	USA (NA)	Old	Benson et al. (1976)
		Pb/Zn (Ag/Au)	EXP	USA (NA)	Old	Beyer et al. (1998)
		Pb/Zn (Ag/Au)	EF	USA (NA)	Old	Blus et al. (1991)
		Pb/Zn (Ag/Au)	EF	USA (NA)	Old	Blus et al. (1999)
		Pb/Zn (Ag/Au)	EF	USA (NA)	Old	Sileo et al. (2001)

(continued)

Table 1 (continued)

Species	N	Metal ore	Biomarker	Country (continent)	Mine status	Reference
Western Canada geese (*Branta canadensis moffitti*)	1	Pb/Zn (Ag/Au)	EF	USA (NA)	Old	Henny et al. (2000)
Wood duck (*Aix sponsa*)	1	Pb/Zn (Ag/Au)	EF	USA (NA)	Old	Blus et al. (1993)
Charadriiformes (1 species)						
Yellow-legged gull (*Larus michahellis*)	1	Pb/Zn/Cu	EXP	Spain (E)	Mining spill	Benito et al. (1999)
Ciconiiformes (1 species)						
White stork (*Ciconia ciconia*)	11	Pb/Zn/Cu	EF	Spain (E)	Mining spill	Baos et al. (2006a)
		Pb/Zn/Cu	EF	Spain (E)	Mining spill	Baos et al. (2012)
		Pb/Zn/Cu	EF	Spain (E)	Mining spill	Pastor et al. (2001)
		Pb/Zn/Cu	EXP	Spain (E)	Mining spill	Rodríguez Álvarez et al. (2013)
		Pb/Zn/Cu	EF	Spain (E)	Mining spill	Smits et al. (2005)
		Pb/Zn/Cu	EF	Spain (E)	Mining spill	Smits et al. (2007)
		Pb/Zn/Cu	EXP	Spain (E)	Mining spill	Benito et al. (1999)
		Pb/Zn/Cu	EF	Spain (E)	Mining spill	Baos et al. (2006b)
		Pb/Zn/Cu	EF	Spain (E)	Mining spill	Baos et al. (2006c)
		Pb/Zn/Cu	EF	Spain (E)	Mining spill	Pastor et al. (2004)
		Pb/Zn/Cu	EXP	Spain (E)	Mining spill	Meharg et al. (2002)
Columbiiformes (2 species)						
Laughing dove (*Spilopelia senegalensis*)	1	Au	EF	Saudi Arabia (AS)	Active	Almalki et al. (2019a)
Mourning doves (*Zenaida macroura*)	1	Pb/Zn	EF	USA (NA)	Old	Beyer et al. (2004)
Coraciiformes (2 species)						
Common Kingfisher (*Alcedo atthis*)	1	Hg	EXP	China (AS)	Old	Abeysinghe et al. (2017)
European bee-eaters (*Merops apiaster*)	1	Cu	EXP	Portugal (E)	Old	Lopes et al. (2010)
Falconiformes (3 species)						
American Kestrel (*Falco sparverius*)	1	Pb/Zn (Ag/Au)	EF	USA (NA)	Old	Henny et al. (1994)
Kestrels (*Falco tinnunculus*)	1	Cu/As	EXP	United Kingdom (E)	Old	Erry et al. (1999a)

Peregrine falcon (*Falco peregrinus*)	1	Pb/Zn/Cu	EXP	Spain (E)	Rodríguez Álvarez et al. (2013)
Galliformes (3 species)					
Free-range chicken	1	Au Mn	EXP	Ghana (AF)	Bortey-Sam et al. (2015)
Northern bobwhites (*Colinus virginianus*)	1	Pb/Zn	EF	USA (NA)	Beyer et al. (2004)
White-tailed ptarmigan (*Lagopus leucurus*)	1	No info	EF	USA (NA)	Larison et al. (2000)
Gruiformes (2 species)					
Eurasian coot (*Fulica atra*)	2	Pb/Zn/Cu	EXP	Spain (E)	Taggart et al. (2006)
		Pb/Zn/Cu	EXP	Spain (E)	Rodríguez Álvarez et al. (2013)
Purple gallinule (*Porphyrio porphyrio*)	1	Pb/Zn/Cu	EF	Spain (E)	Martinez-Haro et al. (2013)
Passeriformes (51 species)					
American dipper (*Cinclus mexicanus*)	2	Hg/Au/Ag	EF	USA (NA)	Henny et al. (2005)
		Cu/Ag/Fe/Pb/Au/Zn	EF	USA (NA)	Strom et al. (2002)
American robins (*Turdus migratorius*)	5	Pb/Zn	EF	USA (NA)	Beyer et al. (2004)
		Pb (Zn/Cu)	EF	USA (NA)	Beyer et al. (2013)
		Pb/Zn (Ag/Au)	EF	USA (NA)	Hansen et al. (2011)
		Pb/Zn (Ag/Au)	EF	USA (NA)	Johnson et al. (1999)
		Pb/Zn (Ag/Au)	EF	USA (NA)	Blus et al. (1995)
American tree sparrow (*Spizella arborea*)	2	Pb/Zn	EF	USA (NA)	Brumbaugh et al. (2010)
		Au	EXP	Canada (NA)	Koch et al. (2005)
Bank swallows (*Riparia riparia*)	2	Pb/Zn	EF	USA (NA)	Beyer et al. (2004)
		Pb (Zn/Cu)	EXP	USA (NA)	Niethammer et al. (1985)
Bhestnut Bulbul (*Hemixos castanonotus*)	1	Hg	EXP	China (AS)	Qiu et al. (2019)
Blue jays (*Cyanocitta cristata*)	1	Pb (Zn/Cu)	EF	USA (NA)	Beyer et al. (2013)
Brown thrasher (*Toxostoma rufum*)	1	Pb/Zn	EF	USA (NA)	Beyer et al. (2004)
Brown-breasted Bulbul	2	Hg	EXP	China (AS)	Qiu et al. (2019)
(*Pycnonotus xanthorrhous*)		Hg	EXP	China (AS)	Abeysinghe et al. (2017)

					Mining spill
					Active
					Old
					Active
					Mining spill
					Mining spill
					Mining spill
					Old
					Old
					Old
					Active/old (whole district)
					Old
					Old
					Old
					Active
					Old
					Old
					Active/old
					Old
					Active/old (whole district)
					Old
					Old
					Old

(continued)

Table 1 (continued)

Species	N	Metal ore	Biomarker	Country (continent)	Mine status	Reference
Brown-cheeked fulvetta (*Alcippe poioicephala*)	1	Hg	EXP	China (AS)	Old	Abeysinghe et al. (2017)
Cactus wren (*Campylorhynchus brunneicapillus*)	1	Ag/Pb/Zn/Cu/Fe Zn(Pb/Cu/Cd/Ag/Au)	EXP	Mexico (NA)	Old (PSG) Active (Charcas)	Monzalvo-Santos et al. (2016)
Canyon towhee (*Melozone fusca*)	1	Ag/Pb/Zn/Cu/Fe Zn(Pb/Cu/Cd/Ag/Au)	EXP	Mexico (NA)	Old (PSG) Active (Charcas)	Monzalvo-Santos et al. (2016)
Carolina wren (*Thryothorus ludovicianus*)	1	Pb (Zn/Cu)	EF	USA (NA)	Active/old (whole district)	Beyer et al. (2013)
Chestnut Bulbul (*Hemixos castanonotus*)	1	Hg	EXP	China (AS)	Old	Abeysinghe et al. (2017)
Chestnut-headed Tesia (*Oligura castaneocoronata*)	2	Hg	EXP	China (AS)	Old	Abeysinghe et al. (2017)
		Hg	EXP	China (AS)	Old	Qiu et al. (2019)
Collared Finchbill (*Spizixos semitorques*)	2	Hg	EXP	China (AS)	Old	Abeysinghe et al. (2017)
		Hg	EXP	China (AS)	Old	Qiu et al. (2019)
Common redpoll (*Carduelis flammea*)	1	Pb/Zn	EF	USA (NA)	Active	Brumbaugh et al. (2010)
Curve-billed thrasher (*Toxostoma curvirostre*)	1	Ag/Pb/Zn/Cu/Fe Zn(Pb/Cu/Cd/Ag/Au)	EXP	Mexico (NA)	Old (PSG) Active (Charcas)	Monzalvo-Santos et al. (2016)
Dark-eyed junco (*Junco hyemalis*)	1	Au	EXP	Canada (NA)	Old	Koch et al. (2005)
Eastern towhees (*Pipilo erythrophthalmus*)	1	Pb (Zn/Cu)	EF	USA (NA)	Active/old (whole district)	Beyer et al. (2013)
Gray jays (*Perisoreus canadensis*)	1	Au	EXP	Canada (NA)	Old	Koch et al. (2005)
Great tit (*Parus major*)	1	Hg	EXP	China (AS)	Old	Abeysinghe et al. (2017)
Grey-cheeked fulvetta (*Alcippe morrisonia*)	1	Hg	EXP	China (AS)	Old	Abeysinghe et al. (2017)
Grey-chinned Minivet (*Pericrocotus solaris*)	1	Hg	EXP	China (AS)	Old	Abeysinghe et al. (2017)
Grey-headed Parrotbill (*Paradoxornis gularis*)	2	Hg	EXP	China (AS)	Old	Abeysinghe et al. (2017)
		Hg	EXP	China (AS)	Old	Qiu et al. (2019)

Species		Metal	EF/EXP	Country	Active/Old	Reference
House sparrow (*Passer domesticus*)	1	Pb/Zn/Ag Cu	EF	Australia (O)	Active (2 sites)	Andrew et al. (2019)
House wren (*Troglodytes aedon*)	2	Au	EXP	USA (NA)	Old	Custer et al. (2002b)
		Au/Ag	EF	USA (NA)	Old	Custer et al. (2007)
Hwamei (*Leucodioptron canorum*)	3	Hg	EXP	China (AS)	Old	Abeysinghe et al. (2017)
		Hg	EXP	China (AS)	Old	Qiu et al. (2019)
		As	EXP	China (AS)	Old	Yang et al. (2018)
Little Pied Flycatcher (*Ficedula westermanni*)	1	Hg	EXP	China (AS)	Old	Abeysinghe et al. (2017)
Light-vented bulbul (*Pycnonotus sinensis*)	1	As	EXP	China (AS)	Old	Yang et al. (2018)
Magpie (*Pica pica*)	1	As	EXP	China (AS)	Old	Yang et al. (2018)
Mountain bluebirds (*Sialia currucoides*)		Cu/Mo	EF	Canada (NA)	Active/old (reclaimed)	O'Brien and Dawson (2016)
Mountain Bulbul (*Ixos mcclellandii*)	2	Hg	EXP	China (AS)	Old	Abeysinghe et al. (2017)
		Hg	EXP	China (AS)	Old	Qiu et al. (2019)
Mountain chickadee (*Poecile gambeli*)	1	Au/Ag	EF	USA (NA)	Old	Custer et al. (2009)
Northern cardinal (*Cardinalis cardinalis*)	2	Pb/Zn	EF	USA (NA)	Old	Beyer et al. (2004)
		Pb (Zn/Cu)	EF	USA (NA)	Old	Beyer et al. (2013)
Northern rough-winged swallows (*Stelgidopteryx serripennis*)	1	Pb (Zn/Cu)	EXP	USA (NA)	Active/old (whole district)	Niethammer et al. (1985)
Pied flycatcher (*Ficedula hypoleuca*)	3	Pb/Zn	EF	Sweden (E)	Old	Berglund et al. (2010)
		Pb/Zn	EXP	Sweden (E)	Old	Berglund 2018
		Pb/Zn	EF	Sweden (E)	Old	Lidman et al. (2020)
Red-billed leiothrix (*Leiothrix lutea*)	1	Hg	EXP	China (AS)	Old	Abeysinghe et al. (2017)
Rough-winged swallows (*Stelgidopteryx ruficollis*)	1	Pb/Zn	EF	USA (NA)	Old	Beyer et al. (2004)
Savannah sparrow (*Passerculus sandwichensis*)	1	Pb/Zn	EF	USA (NA)	Active	Brumbaugh et al. (2010)
Song sparrow (*Melospiza melodia*)	2	Pb/Zn (Ag/Au)	EF	USA (NA)	Old	Hansen et al. (2011)
		Pb/Zn (Ag/Au)	EF	USA (NA)	Old	Johnson et al. (1999)

(continued)

Table 1 (continued)

Species	N	Metal ore	Biomarker	Country (continent)	Mine status	Reference
Southwestern song sparrow (*Melospiza melodia fallax*)	1	Pb/Zn/Cu (Ag/Au)	EF	USA (NA)	Old	Lester and van Riper (2014)
Spot-breasted Scimitar Babbler (*Pomatorhinus mcclellandi*)	2	Hg	EXP	China (AS)	Old	Abeysinghe et al. (2017)
		Hg	EXP	China (AS)	Old	Qiu et al. (2019)
Streak-breasted Scimitar Babbler (*Pomatorhinus ruficollis*)	2	Hg	EXP	China (AS)	Old	Abeysinghe et al. (2017)
		Hg	EXP	China (AS)	Old	Qiu et al. (2019)
Swainson's trushes (*Catharus ustulatus*)	1	Pb/Zn (Ag/Au)	EF	USA (NA)	Old	Hansen et al. (2011)
Tree Sparrow (*Passer montanus*)	3	Hg	EXP	China (AS)	Old	Abeysinghe et al. (2017)
		Hg	EXP	China (AS)	Old	Qiu et al. (2019)
		As	EXP	China (AS)	Old	Yang et al. (2018)
Tree swallow (*Tachycineta bicolor*)	5	Au/Ag/Cu/Zn/Mn/Pb	EF	USA (NA)	Active/old	Custer et al. (2002a)
		Au/Ag	EF	USA (NA)	Old	Custer et al. (2007)
		Au/Ag	EF	USA (NA)	Old	Custer et al. (2009)
		Pb/Zn (Ag/Au)	EF	USA (NA)	Old	Blus et al. (1995)
		Cu/Mo	EF	Canada (NA)	Active/old (reclaimed)	O'Brien and Dawson (2016)
Vinous-throated parrotbill (*Paradoxornis webbianus*)	1	Hg	EXP	China (AS)	Old	Abeysinghe et al. (2017)
Violet-green swallow (*Tachyneta thalassina*)	1	Au/Ag	EF	USA (NA)	Old	Custer et al. (2009)
Western bluebirds (*Sialia mexicana*)	1	Au/Ag	EF	USA (NA)	Old	Custer et al. (2007)
Wood thrush (*Hylocichla mustelina*)	1	Pb (Zn/Cu)	EF	USA (NA)	Active/old (whole district)	Beyer et al. (2013)
Yellow-browed tit (*Sylviparus modestus*)	1	Hg	EXP	China (AS)	Active/old	Abeysinghe et al. (2017)
25 species of passerines		Au/Ag/Cu/Pb/Zn	EXP	Mexico (NA)	Active/old	Chapa-Vargas et al. (2010)
60 species of birds (54 passerines)		Qi: Hg Dia: Sn Dah: Pb/Zn Don: Pb/Zn	EXP	China (AS)	Active/old	He et al. (2020)

Pelecaniformes (9 species)

Black-crowned night-heron (*Nycticorax nycticorax*)	4	Au/Ag	EF	USA (NA)	Old	Henny et al. (2002)
		Au/Ag	EF	USA (NA)	Old	Hoffman et al. (2009)
		Au/Ag	EXP	USA (NA)	Old	Henny et al. (2007)
		Au/Ag	EF	USA (NA)	Old	Hill et al. (2008)
Cattle egret (*Bubulcus ibis*)	1	Pb/Zn/Cu	EXP	Spain (E)	Mining spill	Rodríguez Álvarez et al. (2013)
Glossy ibis (*Plegadis falcinellus*)	2	Pb/Zn/Cu	EXP	Spain (E)	Mining spill	Benito et al. (1999)
		Pb/Zn/Cu	EXP	Spain (E)	Mining spill	Rodríguez Álvarez et al. (2013)
Great Blue herons (*Ardea herodias*)	1	Ag/Pb/Zn (Au/Cu/Cd/Sb)	EXP	USA (NA)	Old	Blus et al. (1985)
Green-backed heron (*Butorides striata*)	1	Pb (Zn/Cu)	EXP	USA (NA)	Active/old	Niethammer et al. (1985)
Grey heron (*Ardea cinerea*)	2	Pb/Zn/Cu	EXP	Spain (E)	Mining spill	Benito et al. (1999)
		Pb/Zn/Cu	EXP	Spain (E)	Mining spill	Rodríguez Álvarez et al. (2013)
Purple heron (*Ardea purpurea*)	1	Pb/Zn/Cu	EXP	Spain (E)	Mining spill	Rodríguez Álvarez et al. (2013)
Snowy egret (*Egretta thula*)	5	Au/Ag	EF	USA (NA)	Old	Henny et al. (2002)
		Au/Ag	EF	USA (NA)	Old	Hoffman et al. (2009)
		Au/Ag	EXP	USA (NA)	Old	Henny et al. (2007)
		Au/Ag	EF	USA (NA)	Old	Hill et al. (2008)
		Au/Ag	EF	USA (NA)	Old	Henny et al. (2017)
Spoonbill (*Platalea leucorodia*)	1	Pb/Zn/Cu	EXP	Spain (E)	Mining spill	Benito et al. (1999)
Phoenicopteriformes (1 species)						
Greater flamingo (*Phoenicopterus ruber*)	1	Pb/Zn/Cu	EXP	Spain (E)	Mining spill	Benito et al. (1999)
Podicipediformes (5 species)						
Black-necked grebes (*Podiceps nigricollis*)	1	Pb/Zn/Cu Ag/Au	EF	Spain (E)	Active/old	Rodríguez-Estival et al. (2019)
Clark's grebes (*Aechmophorus clarkii*)	1	Hg	EF	USA (NA)	Old	Elbert and Anderson (1998)
Great crested grebe (*Podiceps cristatus*)	1	Pb/Zn/Cu	EXP	Spain (E)	Mining spill	Benito et al. (1999)
Red-necked grebes (*Podiceps grisgena*)	1	Hg	EF	Canada (NA)	Old	Weech et al. (2006)

(continued)

Table 1 (continued)

Species	N	Metal ore	Biomarker	Country (continent)	Mine status	Reference
Western grebes (*Aechmophorus occidentalis*)	1	Hg	EF	USA (NA)	Old	Elbert and Anderson (1998)
Strigiformes (5 species)						
Barn owls (*Tyto alba*)	1	Cu/As	EXP	United Kingdom (E)	Old	Erry et al. (1999a)
Eurasian Eagle Owl (*Bubo bubo*)	4	Pb/Zn/Cu/Sn/Fe/Mn/Ag	EF	Spain (E)	Old	Espín et al. (2015)
		Pb/Zn/Cu/Sn/Fe/Mn/Ag	EXP	Spain (E)	Old	Espín et al. (2014b)
		Pb/Zn/Cu/Sn/Fe/Mn/Ag	EF	Spain (E)	Old	Espín et al. (2014a)
		Pb/Zn/Cu/Sn/Fe/Mn/Ag	EF	Spain (E)	Old	Gómez-Ramírez et al. (2011)
Great Horned Owl (*Bubo virginianus*)	1	Pb/Zn (Ag/Au)	EF	USA (NA)	Old	Henny et al. (1994)
Oriental Scops Owl (*Otus sunia*)	1	Hg	EXP	China (AS)	Old	Abeysinghe et al. (2017)
Western Screech-Owl (*Otus kennicotti*)	1	Pb/Zn (Ag/Au)	EF	USA (NA)	Old	Henny et al. (1994)
Suliformes (1 species)						
Double-crested cormorants (*Phalacrocorax auritus*)	1	Au/Ag	EF	USA (NA)	Old	Henny et al. (2002)
Tinamiformes (2 species)						
Darwin's Mothura (*Nothura darwinii*)	1	Past: Au/Ag (Fe/Cu/Zn/Pb/As/Sb) Present: Sn/Zn/Ag/Pb	EF	Bolivia (SA)	Active/old	Garitano-Zavala et al. (2010)
Ornate Tinamou (*Nothoprocta ornata*)	1	Past: Au/Ag (Fe/Cu/Zn/Pb/As/Sb) Present: Sn/Zn/Ag/Pb	EF	Bolivia (SA)	Active/old	Garitano-Zavala et al. (2010)
Numerous species						
14 species of waterfowl		Pb/Zn/Cu	EXP	Spain (E)	Mining spill	Gómez et al. (2004)
16 species of waterfowl		Pb/Zn/Cu	EXP	Spain (E)	Mining spill	Hernández et al. (1999)
195 species of birds		Au	EF	Ghana (AF)	Active	Deikumah et al. (2014)

AMPHIBIA

Anura (6 species)

American bullfrog (*Lithobates catesbeianus*) Named *Rana catesbeiana* in publication.		Pb (Zn/Cu)	EXP	USA (NA)	Active/old	Niethammer et al. (1985)
American toad (*Anaxyrus americanus*) Named *Bufo americanus* in publication		Au	EXP	Canada (NA)	Old	Moriarty et al. (2013)
Asian grass frog (*Fejervarya limnocharis*)		Hg	EXP	China (AS)	Old	Abeysinghe et al. (2017)
Chinese brown frog (*Rana chensinensis*)		Sb	EXP	China (AS)	Active	Fu et al. (2011)
Giant river toad (*Phrynoidis juxtasper*) Named *Bufo juxtasper* in publication.		Cu	EXP	Malaysia (AS)	Active	Lee and Stuebing (1990)
Green frog (*Lithobates clamitans*) Named *Rana clamitans* in publication.		Au	EXP	Canada (NA)	Old	Moriarty et al. (2013)

MAMMALIA

Artiodactyla (11 species)

Caribou (*Rangifer tarandus*)	1	Pb/Zn	EXP	USA (NA)	Active	O'Hara et al. (2003)
Cattle	4	Pb/Zn	EF	Zambia (AF)	Old	Ikenaka et al. (2012)
		Cu	EXP	United Kingdom (E)	Old	Wilson and Pyatt (2007)
		Pb/Zn	EXP	Zambia (AF)	Old	Yabe et al. (2011)
		Pb/Zn	EF	Spain (E)	Old	Rodríguez-Estival et al. (2012)
Chamois (*Rupicapra rupicapra*)	1	Pb	EXP	Slovenia (E)	Active	Doganoc and Gačnik (1995)
Goat	3	Au Mn	EXP	Ghana (AF)	Active	Bortey-Sam et al. (2015)
		Au	EXP	Nigeria (AF)	Active	Jubril et al. (2017)
		Au	EF	Nigeria (AF)	Active	Jubril et al. (2019)
Hippopotamus (*Hippopotamus amphibious L.*)	1	Cu/Co	EXP	Zambia (AF)	Active	Mwase et al. (2002)
Kafue lechwe (*Kobus leche kafuensis*)	2	Cu/Co	EXP	Zambia (AF)	Active	Syakalima et al. (2001b)
			EXP	Zambia (AF)	Active	Syakalima et al. (2001a)

(continued)

Table 1 (continued)

Species	N	Metal ore	Biomarker	Country (continent)	Mine status	Reference
Red deer (*Cervus elaphus*)	12	Hg/Pb	EF	Spain (E)	Old	Berzas Nevado et al. (2012)
		Pb/Zn	EF	Spain (E)	Old	Castellanos et al. (2010)
		Pb/Zn	EF	Spain (E)	Old	Castellanos et al. (2015)
		Pb	EXP	Slovenia (E)	Active	Doganoc and Gačnik (1995)
		Pb/Zn	EXP	Spain (E)	Old	Reglero et al. (2008)
		Pb/Zn	EF	Spain (E)	Old	Reglero et al. (2009a)
		Pb/Zn	EXP	Spain (E)	Old	Taggart et al. (2011)
		Pb/Zn	EF	Spain (E)	Old	Reglero et al. (2009b)
		Pb/Zn	EF	Spain (E)	Old	Rodríguez-Estival et al. (2011a)
		Pb/Zn	EF	Spain (E)	Old	Rodríguez-Estival et al. (2011b)
		Pb/Zn	EF	Spain (E)	Old	Rodríguez-Estival et al. (2013a)
		Pb/Zn	EF	Spain (E)	Old	Rodríguez-Estival et al. (2013b)
Roe deer (*Capreolus capreolus*)	1	Pb	EXP	Slovenia (E)	Active	Doganoc and Gačnik (1995)
Sheep	4	Au Mn	EXP	Ghana (AF)	Active	Bortey-Sam et al. (2015)
		Pb/Zn	EXP	Spain (E)	Old	Pareja-Carrera et al. (2014)
		Pb/Zn	EF	Spain (E)	Old	Rodríguez-Estival et al. (2012)
		Pb/Zn/Cu	EXP	United Kingdom (E)	Old	Smith et al. (2010)
White-tailed deer (*Odocoileus virginianus*)	2	Pb (Zn/Cu/Ag)	EXP	USA (NA)	Old (remediated)	Beyer et al. (2007)
		Pb/Zn	EXP	USA (NA)	Old	Conder and Lanno (1999)
Wild boar (*Sus scrofa*)	8	Hg (Pb)	EF	Spain (E)	Old	Berzas Nevado et al. (2012)
		Pb	EXP	Slovenia (E)	Active	Doganoc and Gačnik (1995)
		Pb/Zn	EF	Spain (E)	Old	Reglero et al. (2009b)
		Pb/Zn	EF	Spain (E)	Old	Rodríguez-Estival et al. (2011a)
		Pb/Zn	EF	Spain (E)	Old	Rodríguez-Estival et al. (2011b)

		Pb/Zn	EF	Spain (E)	Old	Rodríguez-Estival et al. (2013a)
		Pb/Zn	EF	Spain (E)	Old	Rodríguez-Estival et al. (2013b)
		Pb/Zn	EXP	Spain (E)	Old	Taggart et al. (2011)
Carnivora (8 species)						
Black bears (*Ursus americanus*)	1	Au, Ag, Cu and Zn	EXP	USA (NA)	Old	Peplow and Edmonds (2005)
Common genets (*Genetta genetta*)	1	Az: Pb/Zn/Cu Al: Pb/Zn/Hg	EXP	Spain (E)	Old/Mining spill	Millán et al. (2008)
Egyptian mongoose (*Herpestes ichneumon*)	1	Az: Pb/Zn/Cu Al: Pb/Zn/Hg	EXP	Spain (E)	Old/Mining spill	Millán et al. (2008)
Eurasian badger (*Meles meles*)	1	Az: Pb/Zn/Cu Al: Pb/Zn/Hg	EXP	Spain (E)	Old/Mining spill	Millán et al. (2008)
Eurasian otter (*Lutra lutra*)	2	Pb/Zn/Cu	EXP	Spain (E)	Mining spill	Delibes et al. (2009)
		Pb Hg	EXP	Spain (E)	Old	Rodríguez-Estival et al. (2020)
Iberian Lynx (*Lynx pardinus*)	1	Az: Pb/Zn/Cu Al: Pb/Zn/Hg	EXP	Spain (E)	Old/Mining spill	Millán et al. (2008)
Mink (*Mustela vison*)	2	Pb/Zn (Ag/Au)	EXP	USA (NA)	Old	Blus et al. (1987)
		Pb/Zn (Ag/Au)	EXP	USA (NA)	Old	Blus and Henny (1990)
Red fox (*Vulpes vulpes*)	1	Az: Pb/Zn/Cu Al: Pb/Zn/Hg	EXP	Spain (E)	Old/Mining spill	Millán et al. (2008)
Dasyuromorphia (1 species)						
Northern quoll (*Dasyurus hallucatus*)	2	Mn	EF	Australia (AU)	Active	Amir Abdul Nasir et al. (2018a)
		Mn	EXP	Australia (AU)	Active	Amir Abdul Nasir et al. (2018b)
Eulipotyphla (4 species)						
Eurasian shrew (*Sorex araneus*)	3	Pb/Zn	EXP	United Kingdom (E)	Old	Roberts and Johnson (1978)
		No info	EXP	United Kingdom (E)	Old (revegetated)	Andrews et al. (1984)
		Pb/Zn	EF	United Kingdom (E)	Old	Roberts et al. (1978)
Elliot's short-tailed shrew (*Blarina hylophaga*)	1	Pb/Zn	EF	USA (NA)	Old	Phelps and McBee (2009)

(continued)

Table 1 (continued)

Species	N	Metal ore	Biomarker	Country (continent)	Mine status	Reference
Greater white-toothed shrew (*Crocidura russula*)	6	Pb/Zn/Cu	EF	Spain (E)	Mining spill	Sánchez-Chardi et al. (2009)
		Pb/Zn/Cu	EXP	Spain (E)	Mining spill	Sánchez-Chardi (2007)
		Au/Ag/As	EF	France (E)	Old (Partially remediated)	Drouhot et al. (2014)
		Pb/Zn	EF	Portugal (E)	Old	Marques et al. (2007)
		Pb/Zn/Cu/Fe	EF	Portugal (E)	Old	Sánchez-Chardi et al. (2007b)
		Pb/Zn/Cu/Fe	EF	Portugal (E)	Old	Sánchez-Chardi et al. (2008)
Cinereus shrew (*Sorex cinereus*)	1	Au	EXP	Canada (NA)	Old	Moriarty et al. (2012)
Lagomorpha (3 species)						
Artic hares (*Lepus arcticus*)	1	Pb/Zn	EF	Canada (NA)	Old (reclamation)	Amuno et al. (2016)
European rabbit (*Oryctolagus cuniculus*)	1	Pb/Zn/Cu/Sn/Fe/Mn/Ag	EXP	Spain (E)	Old	Espín et al. (2014b)
Snowshoe hare (*Lepus americanus*)	2	Au	EF	Canada (NA)	Old (Remediation Project)	Amuno et al. (2018a)
		Au	EF	Canada (NA)	Old (Remediation Project)	Amuno et al. (2018b)
Perissodactyla (1 species)						
Horse (livestock)	1	Pb/Zn/Cu	EXP	Spain (E)	Mining spill	Madejón et al. (2009)
Primates (2 species)						
Eastern wooly lemur (*Avahi laniger*)		Ni/Co	EF	Madagascar (AF)	Active	Junge et al. (2017)
Greater sportive lemur (*Lepilemur mustelinus*)		Ni/Co	EF	Madagascar (AF)	Active	Junge et al. (2017)
Rodentia (39 species)						
Algerian mouse (*Mus spretus*)	13	Pb/Zn/Cu	EF	Spain (E)	Mining spill	Bonilla-Valverde et al. (2004)
		Au/Ag/As	EF	France (E)	Old (Partially remediated)	Drouhot et al. (2014)

		Pb/Zn/Cu	EF	Spain (E)	Mining spill	Festa et al. (2003)
		Pb/Zn/Cu/Fe	EF	Portugal (E)	Old	Marques et al. (2008)
		Pb/Zn/Cu Ag/Au	EF	Spain (E)	Past/present	Mateos et al. (2008)
		Sn/Cu	EF	Portugal (E)	Active	Nunes et al. (2001a)
		Sn/Cu	EF	Portugal (E)	Active	Nunes et al. (2001b)
		Cu	EF	Portugal (E)	Old	Pereira et al. (2006)
		Al: Pb/Zn/Cu/Fe P: Pb/Zn	EF	Portugal (E)	Old	Quina et al. (2019)
		Pb/Zn/Cu	EF	Spain (E)	Mining spill	Ruiz-Laguna et al. (2001)
		Pb/Zn/Cu	EF	Spain (E)	Mining spill	Tanzarella et al. (2001)
		Pb/Zn/Cu	EF	Spain (E)	Mining spill	Udroiu et al. 2008
		Sn/Cu	EF	Portugal (E)	Active	Viegas-Crespo et al. (2003)
Baja pocket mouse (*Chaetodipus rudinoris*)	1	Cu Au/Ag	EXP	Mexico (NA)	Active (S. Rosalia) Old (El Triunfo)	Méndez-Rodríguez and Álvarez-Castañeda (2019)
Balochistan gerbil (*Gerbillus nanus*)	1	Au	EF	Saudi Arabia (AS)	Active	Almalki et al. (2019b)
Bank vole (*Clethrionomys glareolus*)	7	Pb/Zn	EXP	United Kingdom (E)	Old	Johnson et al. (1978)
		Pb/Zn	EXP	Ireland (E)	Old	Purcell et al. (1992)
		Cu/As Pb/Ag/Zn	EXP	United Kingdom (E)	Old	Erry et al. (1999b)
		Cu/As Pb/Ag/Zn	EXP	United Kingdom (E)	Old (Revegetated)	Erry et al. (2000)
		Cu/As Pb/Ag/Zn	EXP	United Kingdom (E)	Old (Revegetated)	Erry et al. (2005)
		Pb	EXP	United Kingdom (E)	Old	Milton et al. (2003)
		Pb/Zn	EF	United Kingdom (E)	Old	Roberts et al. (1978)
Common vole (*Microtus arvalis*)	1	Au/Ag/As	EF	France (E)	Old (Partially remediated)	Drouhot et al. (2014)

(continued)

Table 1 (continued)

Species	N	Metal ore	Biomarker	Country (continent)	Mine status	Reference
Deer mouse (*Peromyscus maniculatus*)	8	Pb Zn Cd	EF	USA (NA)	Old (Remediated)	Coolon et al. (2010)
		Cu	EXP	Canada (NA)	Old	Laurinolli and Bendell-Young (1996)
		Au	EXP	Canada (NA)	Old	Ollson et al. (2009)
		Ag/Au	EXP	USA (NA)	Old	Pascoe et al. (1994)
		Pb/Zn (Ag/Au)	EXP	USA (NA)	Old	Blus et al. (1987), Blus et al. (1987)
		Pb/Zn	EF	USA (NA)	Old	Phelps and McBee (2009)
		Au	EXP	Canada (NA)	Old	Saunders et al. (2011)
		Zn/Cu	EXP	Canada/USA (NA)	Active	Smith and Rongstad (1982)
Dulzura kangaroo rat (*Dipodomys simulans*)	1	Cu Au/Ag	EXP	Mexico (NA)	Active (S. Rosalia) Old (El Triunfo)	Méndez-Rodríguez and Álvarez-Castañeda (2019)
Dusky rice rat (*Melanomys caliginosus*)	1	Au	EXP	Colombia (SA)	Active	Sierra-Marquez et al. (2018)
Eastern woodrat (*Neotoma floridana*)	1	Pb/Zn	EF	USA (NA)	Old	Phelps and McBee (2009)
Field vole (*Microtus agrestis*)	4	No info	EXP	United Kingdom (E)	Old (revegetated)	Andrews et al. (1984)
		Pb/Zn	EF	United Kingdom (E)	Old	Roberts et al. (1978)
		Pb/Zn	EXP	United Kingdom (E)	Old	Johnson et al. (1978)
		Pb/Zn	EXP	United Kingdom (E)	Old	Roberts and Johnson (1978)
Forest small rice rat (*Microryzomys minutus*)	1	Au	EXP	Colombia (SA)	Active	Sierra-Marquez et al. (2018)
Fulvous harvest mouse (*Reithrodontomys fulvescens*)	1	Pb/Zn	EF	USA (NA)	Old	Phelps and McBee (2009)
Hispid cotton rat (*Sigmodon hispidus*)	1	Pb/Zn	EF	USA (NA)	Old	Phelps and McBee (2009)
House mouse (*Mus musculus*)	1	Pb/Zn	EF	USA (NA)	Old	Phelps and McBee (2009)

Species		Metal	Study type	Country (Continent)	Mine status	Reference
Little desert pocket mice (*Chaetodipus arenarius*)	1	Au/Ag	EXP	Mexico (NA)	Old	Méndez-Rodríguez and Álvarez-Castañeda (2016)
Meadow vole (*Microtus pennsylvanicus*)	5	Au	EXP	Canada (NA)	Old	Saunders et al. (2011)
		Zn/Cu	EXP	Canada/USA (NA)	Active	Smith and Rongstad (1982)
		Au	EF	Canada (NA)	Old	Saunders et al. (2009)
		Au	EF	Canada (NA)	Old	Saunders et al. (2010)
		Ag/Au	EF	USA (NA)	Old	Pascoe et al. (1994)
Merriam's kangaroo rat (*Dipodomys merriami*)	3	Pb/Zn/Ag (Cu/Au)	EF	Mexico (NA)	Active/old	Espinosa-Reyes et al. (2010)
		Pb/Zn/Ag (Cu/Au)	EF	Mexico (NA)	Active	Espinosa-Reyes et al. (2014)
		Pb/Zn/Ag (Cu/Au)	EF	Mexico (NA)	Active/old	Jasso-Pineda et al. (2007)
Mexican harvest mouse (*Reithrodontomys mexicanus*)	1	Au	EXP	Colombia (SA)	Active	Sierra-Marquez et al. (2018)
Muskrat (*Ondatra zibethicus*)	2	Pb/Zn (Ag/Au)	EXP	USA (NA)	Old	Blus et al. (1987)
		Pb (Zn/Cu)	EXP	USA (NA)	Active/old	Niethammer et al. (1985)
Nelson's pocket mouse (*Chaetodipus nelsoni*)	2	Pb/Zn/Ag (Cu/Au)	EF	Mexico (NA)	Active	Espinosa-Reyes et al. (2014)
		Pb/Zn/Ag (Cu/Au)	EF	Mexico (NA)	Active/old	Jasso-Pineda et al. (2007)
Nephelomys pectoralis	1	Au	EXP	Colombia (SA)	Active	Sierra-Marquez et al. (2018)
Northern red-backed vole (*Myodes rutilus*)	1	Pb/Zn	EF	USA (NA)	Active	Brumbaugh et al. (2010)
Persian jird (*Meriones persicus*)	2	Cu	EXP	Iran (AS)	Active (initial stage)	Khazaee et al. (2016)
		Fe	EF	Iran (AS)	Active	Shahsavari et al. (2019)
Plains harvest mouse (*Reithrodontomys montanus*)	1	Pb/Zn	EF	USA (NA)	Old	Phelps and McBee (2009)
Plateau mice (*Peromyscus melanophrys*)	2	Pb/Zn/Ag	EF	Mexico (NA)	Old	Mussali-Galante et al. (2013a)
		Pb/Zn/Ag	EF	Mexico (NA)	Old	Tovar-Sánchez et al. (2012)
Prairie vole (*Microtus ochrogaster*)	1	Pb/Zn	EF	USA (NA)	Old	Phelps and McBee (2009)
Rice Field Rat (*Rattus argentiventer*)	1	Hg	EXP	China (AS)	Old	Abeysinghe et al. (2017)
Silky Oldfield mouse (*Thomasomys bombycinus*)	1	Au	EXP	Colombia (SA)	Active	Sierra-Marquez et al. (2018)

(continued)

Table 1 (continued)

Species	N	Metal ore	Biomarker	Country (continent)	Mine status	Reference
Southern pygmy mice (*Baiomys musculus*)	1	Pb/Zn/Ag	EF	Mexico (NA)	Old	Tovar-Sánchez et al. (2012)
Southern spiny pocket mouse (*Heteromys australis*)	1	Au	EXP	Colombia (SA)	Active	Sierra-Marquez et al. (2018)
Spiny pocket mouse (*Chaetodipus spinatus*)	2	Au/Ag	EXP	Mexico (NA)	Old	Méndez-Rodríguez and Alvarez-Castañeda (2016)
		Cu Au/Ag	EXP	Mexico (NA)	Active (S. Rosalia) Old (El Triunfo)	Méndez-Rodríguez and Álvarez-Castañeda (2019)
Thomasomys nicefori	1	Au	EXP	Colombia (SA)	Active	Sierra-Marquez et al. (2018)
Tundra vole (*Microtus oeconomus*)	1	Pb/Zn	EF	USA (NA)	Active	Brumbaugh et al. (2010)
Voles (*Microtus spp.*)	1	Pb/Zn (Ag/Au)	EXP	USA (NA)	Old	Blus et al. (1987)
White-footed mouse (*Peromyscus leucopus*)	4	Pb/Zn	EF	USA (NA)	Old	Phelps and Mcbee (2010)
		Pb	EF	USA (NA)	Old	Beyer et al. (2018)
		Pb/Zn/Cd	EF	USA (NA)	Old (Remediated)	Coolon et al. (2010)
		Pb/Zn	EF	USA (NA)	Old	Phelps and McBee (2009)
Brown rat (*Rattus norvegicus*)	1	Au Mn	EXP	Ghana (AF)	Active	Bortey-Sam et al. (2016)
Wild rats (*Rattus sp*)	4	Au Mn			Active	Bortey-Sam et al. (2016)
		Cu	EF	Portugal (E)	Old	Pereira et al. (2006)
		Pb/Zn	EF	Zambia (AF)	Old	Nakayama et al. (2011)
		Kabwe Pb/Zn Chingola Cu/Co	EF	Zambia (AF)	Active/old	Nakayama et al. (2013)
Wood mouse (*Apodemus sylvaticus*)	12	M1: Fe M2: Pb/Ag C1: Zn/Pb/Ag	EF	France (E)	Old	Camizuli et al. (2018)

		Pb	EXP	Ireland and United Kingdom (E)	Old	Milton et al. (2002)
		Pb/Zn	EXP	Ireland and United Kingdom (E)	Active/old	Milton et al. (2004)
		Pb/Zn	EXP	Ireland (E)	Old	Milton and Johnson (1999)
		Cu/As Pb/Ag/Zn	EXP	United Kingdom (E)	Old	Erry et al. (1999b)
		Cu/As Pb/Ag/Zn	EXP	United Kingdom (E)	Old (Revegetated)	Erry et al. (2000)
		Cu/As Pb/Ag/Zn	EXP	United Kingdom (E)	Old (Revegetated)	Erry et al. (2005)
		Au/Ag/As	EF	France (E)	Old (Partially remediated)	Drouhot et al. (2014)
		Pb/Zn	EF	United Kingdom (E)	Old	Roberts et al. (1978)
		Pb/Zn	EXP	United Kingdom (E)	Old	Roberts and Johnson (1978)
		Pb/Zn	EXP	Ireland (E)	Old	Purcell et al. (1992)
		Pb/Zn	EXP	United Kingdom (E)	Old	Johnson et al. (1978)
Woodland vole (*Microtus pinetorum*)	1	Pb/Zn	EF	USA (NA)	Old	Phelps and McBee (2009)
REPTILIA						
Crocodilia (1 species)						
Crocodiles (*Crocodylus niloticus*)	1	Cu/Co	EXP	Zambia (AF)	Active	Almli et al. (2005)
Squamata (6 species)						
Big-eyed rat snake (*Ptyas dhumnades*)	1	Hg	EXP	China (AS)	Old	Abeysinghe et al. (2017)
Common lizard (*Lacerta vivipara*)	1	Pb	EXP	United Kingdom (E)	Old	Avery et al. (1983)
Giant girdled or sungazer lizard (*Smaug giganteus*)	1	Au	EF	South Africa (AF)	Active	McIntyre and Whiting (2012)
Large psammodromus (*Psammodromus algirus*)	1	Pb/Zn/Cu	EXP	Spain (E)	Mining spill	Márquez-Ferrando et al. (2009)
Moorish wall gecko (*Tarentola mauritanica*)	1	Pb/Zn/Cu	EXP	Spain (E)	Mining spill	Fletcher et al. (2006)

(continued)

Table 1 (continued)

Species	N	Metal ore	Biomarker	Country (continent)	Mine status	Reference
Northern water snake (*Nerodia sipedon*)	1	Pb	EXP	USA (NA)	Active/old	Niethammer et al. (1985)
Testudines (2 species)						
Mediterranean Pond Turtle (*Mauremys leprosa*)	2	Pb/Ag/Zn	EXP	Spain	Old	Martínez-López et al. (2017)
		Pb Hg	EF	Spain	Old	Ortiz-Santaliestra et al. (2019)
Red-eared slider turtle (*Trachemys scripta*)	2	Pb/Zn	EF	USA (NA)	Old	Hays and McBee (2007)
		Pb/Zn	EF	USA (NA)	Old	Hays and McBee (2010)

N = number of studies reporting that species; Country and Continent (E = Europe; NA = North America; SA = South America; AS = Asia; AU = Australia; AF = Africa); Main metal ore extracted (secondary in brackets); Type of biomarker used (EF = Effect; EXP = Exposure); and Reference

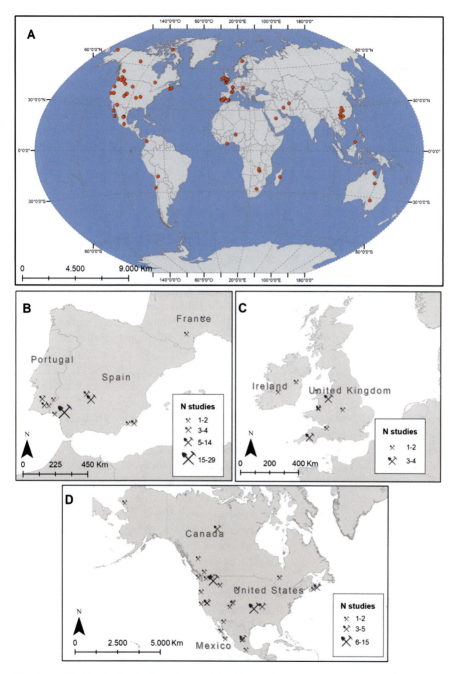

Fig. 1 (**a**) Worldwide distribution of terrestrial mining pollution studies, represented as red dots; (**b–d**) Details of zones with higher densities of studies (Spain, United Kingdom, North America, respectively)

total remediation, revegetation, or reclamation activities. Around 52% of the studies in old mines were done in North America, including studies in Mexico, Canada, and especially in the United States of America (USA), where 45 studies were conducted. One of the most contaminated and studied mining sites was the Coeur d'Alene (CDA) River Basin in northern Idaho, where studies reported waterfowl dying since, at least, the beginning of the twentieth century. On the other hand, only two studies took place in South America, specifically in Bolivia and Chile. The other continent that gathered a great number of studies in derelict mines was Europe (40%). A European area in which mining has historically been outstanding is the Iberian Peninsula, which encompasses Spain and Portugal. 24% of the studies about old mines were developed there, where the Iberian Pyrite Belt (IPB), one of the largest metallogenetic provinces of massive sulfide deposits in the world, has been already exploited 5,000 years ago (Nocete et al. 2005). Also in Spain is the Almadén Mining District, the largest Hg mining area in the world until the beginning of the twenty-first century (Rodríguez-Estival et al. 2020). The United Kingdom is another European country with a legacy of historical abandoned mining sites (11% of old mines), with hundreds of mines peaking during the nineteenth century which have left wastes and tailings (Milton et al. 2003). In Asia, the five publications about derelict mines were located in China. In Africa there were also studies on closed mines (3%), all of them in Zambia. There, the Kabwe mine has been ranked among the top ten most polluted places in the world, with highly contaminated soils, even affecting children (Nakayama et al. 2011).

The number of studies developed in active mines was much lower than that of former mines (30 studies at active mines -16%- and 16 that were both at active and former mines, Fig. 2). Some of them had a long working history, but unlike the abandoned ones, they were at an operational stage when research was carried out. Fourteen studies on actives mines took place in North America. An historical mining area still working in the USA is the Southeast Missouri Lead Mining District. Although the Old Lead Belt was abandoned in 1972, the New Lead Belt has been continuing since 1955 till today. In South America there was one study in Bolivia and another in Colombia. The number of studies on active mines in Europe corresponded to 8 publications, three of them in Portugal. In fact, the number of studies referring to active mining areas in Africa was higher than in Europe (12 papers -27%-). There, several countries are currently extracting metals, such as Zambia, in which Cu accounts for 60% of exports (Ikenaka et al. 2012), but also Ghana, Madagascar, Nigeria, or South Africa. The study of this pollution in developing countries such as Zambia and Ghana, with an extremely active mining industry, is very important. Seven publications were carried out in Asian countries, including China, Iran, Malaysia, and Saudi Arabia. Finally, only three publications were carried out in Australia. While the number of pollution assessments in operational mines developed in Europe was apparently low, mining production in Europe has decreased in the last 20 years (Reichl and Schatz 2019). Thus, it is not surprising that we found a lower number of studies in operational mines in Europe than in Africa or Asia, which are regions with increasing production rates in the last 20 years, with China as the world's largest producer of several metals (Reichl and

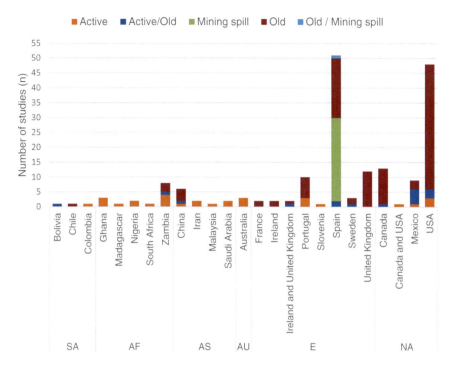

Fig. 2 Studies of metalliferous mining in terrestrial and semi-terrestrial vertebrates by country and continent

Schatz 2019). In fact, we consider that the number of studies found both in China and Australia, two countries ranked among the four biggest mining nations (Reichl and Schatz 2019), was lower than expected.

Some of the papers compiled for the present review quoted some pond dam breaches, such as that occurred in 1977 in Big River (USA) (Niethammer et al. 1985) or the breach of the Milltown Dam upstream of Missoula in the Upper Clark Fork River Basin (USA) (Langner et al. 2012). Other studies reported accidents but give scarce information about its effects on wildlife, such as in Baia Mare and Baia Borsa (Romania), where dam failures occurred in January and March 2000 (Koenig 2000; Macklin et al. 2003). These accidents released a great amount of cyanide and contaminant metals. Macklin et al. (2003) only reported some fish deaths, and Koenig (2000) stated that two fish species found only in the upper Tisza may had been pushed to the brink of extinction, and mentioned the death of an endangered white-tailed eagle and the possibility of an effect over otter populations. But these were only mentions, without any research data or explanations in relation to it. The only exception that we found in which specific risk assessment studies of a mining accident in terrestrial and semi-terrestrial wildlife were developed and published in peer-reviewed journals is the Aznalcóllar mining spill, which took place in April 1998. The 29 studies about the effects on the ecosystem of mining spillages were all related to Aznalcóllar (Fig. 2). Its importance comes from the magnitude of the

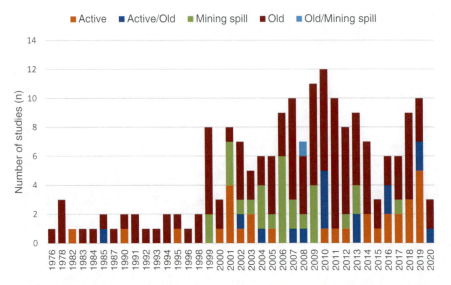

Fig. 3 Studies of metalliferous mining in terrestrial and semi-terrestrial vertebrates per year (1976–2020*). *Studies from 2020 were already available online at the end of 2019

spillage but also because it took place in the surrounding of Doñana National Park, the largest reserve of bird species in Europe, partly protected as a World Heritage Site, Biosphere Reserve and Ramsar Site. Readily after the event and in the following years (Fig. 3), several studies were accomplished in and around Doñana National and Natural Park and along Guadiamar River. Those studies assessed metallic pollution of sediments and water as well as bioaccumulation and effects on wildlife, including different species of birds, small and big mammals or reptiles (Table 1), but also humans (Gil et al. 2006). Such research produced great number of papers on the topic and showed the consequences that an accident of this magnitude could have. As stated before, it is highly remarkable that all studies about effects of a mining spillage on terrestrial vertebrates were from the Aznalcóllar mining accident. Despite several mining accidents have taken place throughout the world, there are only some studies in plants or aquatic organisms. This is possibly due to the fact that mine spill usually consists in discharges of wastes in aquatic environment, thus ignoring the connection between aquatic and terrestrial environment.

In general, studies regarding exclusively metal ore mining pollution represent a small proportion of studies in the field of metal pollution on fauna, particularly in terrestrial and semi-terrestrial vertebrates. This is in line with a recent review of metal and metalloid pollution studies on wildlife species in Latin America (Di Marzio et al. 2019), in which terrestrial ecosystems represent only 4% of the studies. Moreover, as it occurs in reports of tailings dams (Rico et al. 2008), there is a lack of studies and an uneven geographical distribution of mine exploitations and ecotoxicological studies. Such is the case of Chile and Peru, the two leading producers of Cu (Menzie et al. 2013; Reichl and Schatz 2019), or the previously stated example of Australia and China, where ecotoxicological studies are scarce. The majority of studies found were carried out in old mines, with a poor development of research in areas where mining is still an ongoing industry. We hypothesized two main reasons for this difference. First, nowadays legislation regarding operational mines and subsequent mine closure is more developed, starting in most countries in the 1970s (UNDP and UN Environment 2018). For this reason, mining pollution originated in operational mines should be under monitoring by governments and mining companies. However, in some developing countries and politically unstable countries this legislation is usually lax or even nonexistent (UNDP and UN Environment 2018), as in some Southern Mediterranean countries, where mining impacts and risks on health and environment may not yet have been identified as a social and public concern (Doumas et al. 2018). In addition, it is hard to believe that in countries where humans are being directly affected by mining pollution but, even so, disregarded (such us indigenous people in Bolivia, Peru or Brazil, Sault 2018; IRP 2020), research would be focused on wildlife. The second reason is that derelict mines could really represent a bigger problem for humans and environment health than active mine sites. Abandoned mine sites represent a great environmental problem at global scale. The estimated number of more than one million derelict mines (Venkateswarlu et al. 2016) is hardly ascertainable. Developed countries have only recently started to develop programs and legislation regarding the inventory and management of abandoned mine sites and their wastes. Since the last quarter of the twentieth century there has been a significant development of legislation and some actions started to deal with industrial legacy through clean-up, remediation, and rehabilitation projects (Carvalho 2017). These remediation actions started in the USA with the Superfund project in 1980. Also, the European Union published in 2006 the Directive 2006/21/EC, dedicated to the management of mining waste from extractive industries. Canada created in 2002 the National Orphaned/Abandoned Mines Initiative, and Australia passed the mining rehabilitation Fund Act in 2012. After these projects and laws, countries have started to estimate the amount of abandoned mines and waste facilities. It is estimated that in Australia there are over 50,000 abandoned mines (as of 2012), in Canada over 10,000 (as of 2000), and in South Africa nearly 6,000 (as of 2009) (UNDP and UN Environment 2018). In Europe, information from 18 national inventories showed the existence of over 3,400 closed and abandoned waste facilities (as of 2017), 1,234 of them related to metallic minerals and precious metals (European Commission 2017). In the USA, the Bureau of Land Management has estimated as many as 500,000 abandoned mines, but most have not even been mapped (Bureau of Land Management 2019). Although countries such as Australia have already started to rehabilitate some of these abandoned mine lands (Cristescu et al. 2012), the extent of the problem is not only the huge number of abandoned wastes, but also the economic cost of their rehabilitation which must be covered by the governments. For example, in Queensland, Australia, the cost of rehabilitating 15,000 abandoned mines is estimated at AU1$ billion (UNDP and UN Environment 2018), and the U.S. Department of Interior estimated that cleaning up degraded land and streams on more than 500,000 abandoned rock mining sites will cost US$32–72 billion (Tyler Miller and Spoolman

2009). Hopefully, as some of these legislation and projects are very recent, in the next years we will see an increment in the study and management of these areas. Regarding the studies included in this review, they should be useful for human and environmental health risk assessment in these derelict mining sites. Information about metals exposure and effects in sentinel species must be included in reclamation and rehabilitation protocols. Sometimes, these plans are only based on recolonization of flora and fauna, and species richness and densities (Cristescu et al. 2012). However, information on remaining metals bioavailable to wildlife and humans, as well as their effects, is essential and must be included (Camizuli et al. 2018). In addition to include parameters of metal exposure and effect as part of the protocols, the already published literature included in this review may serve as a baseline for further developing legislation on the topic.

3.2 Main Metal Ore

Regarding the main metals extracted on the reviewed studies, there was a great diversity in the main polymetallic ores (Table 1). However, Pb/Zn were the most common as main metals extracted (49 studies) and this number increased up to 105 if we included Pb/Zn polymetallic ores extracted with other metals as silver (Ag), gold (Au), manganese (Mn), tin (Sn), Cu, and Fe. These deposits are a common occurrence worldwide (Gutiérrez et al. 2016) as showed by the range of studies in different countries with Pb/Zn extraction included in this review. As was extracted as main ore, or in co-occurrence with Cu or Ag/Au, in 6 mine sites, located in United Kingdom (4), China (1), and France (1). The extraction of Au as major metal occurred in 17 mine sites, some of them as small-scale or artisanal mining, as in Nigeria. But, in addition to the previously mentioned co-occurrence with Pb/Zn, Au frequently appeared as main metal ore extracted together with Ag (10 sites, mainly in the USA), and occasionally with Ag and Hg, Cu, or As. Regarding Hg, it was extracted as major metal in 10 sites, all of them old mine sites, as Hg is nowadays less extracted than other metals (Reichl and Schatz 2019). Hg was not extracted together with any other metal, but some studies included reports of two nearby mining sites: a Hg mine site and other metal mine, as in Almadén Mining District (Hg) and Sierra Madrona and Alcudia Valley (Pb) in Spain (Ortiz-Santaliestra et al. 2019; Rodríguez-Estival et al. 2020). Studies related to Cu as main metal extracted were 7, comprising several countries. Its co-occurrence was mainly with Sn, cobalt (Co), As, or molybdenum (Mo). Finally, an antimony (Sb) mine was studied in China and an Mn mine in Australia.

3.3 Sentinel Species

In the reviewed literature, numerous species have been used as sentinel species to assess the exposure and the potential effects of metalliferous mining pollution, although some groups were widely more studied than others (Table 1). Most studies focused on a single species, but some studies included more than one species belonging to the same or different taxa group in their risk assessment. Birds appeared in 78 of the papers reviewed, including at least 114 different species. The value of birds as sentinels of environmental pollution has been broadly recognized and they have a long history in ecotoxicological research (Gómez-Ramírez et al. 2014). In particular, raptors were widely used as sentinel species, including 17 species of Accipitriformes, Falconiformes, and Strigiformes (Table 1). Eurasian eagle owl (*Bubo bubo*) and black kite (*Milvus migrans*) were the raptors more frequently used, although all studies for each species were at the same mine site. Some characteristics that make raptors good sentinels include their trophic position at the top of the food chain, their relatively long lifespan over which to accumulate contaminants, and their integration of exposure both over time and relatively large spatial areas (Gómez-Ramírez et al. 2014). However, birds of prey often forage over large territories, which is an important factor when assessing a local source of pollution. For this reason, nestlings are usually captured instead of adults, as they are fed entirely on food resources obtained locally (Baos et al. 2006c). Waterfowl and wading birds were also used as sentinels, especially near lakes, rivers, or wastewater ponds, including 36 different species (Table 1). We included them in the review as they are also connected to terrestrial ecosystems. Also, they have been severely affected by mining pollution, with swans and geese found dead in Coeur D'Alene River Basin and Tri-State Mining District (Blus et al. 1991; Sileo et al. 2001; Carpenter et al. 2004). Waterfowl and wading birds are usually regarded as good bioindicators of metals, especially species easily recognizable, long-lived, abundant and that have a wide geographical range (Kalisińska et al. 2004). In addition, as ingestion is the primary pathway of contaminant exposure in birds (Smith et al. 2007) and this groups of birds mainly feed on plants, invertebrates, and fish, ingestion of metal through polluted food is very likely. Canada goose (*Branta Canadensis*), mallard (*Anas platyrhynchos*), tundra swan (*Cygnus columbianus*), white stork (*Ciconia ciconia*), snowy egrets (*Egretta thula*), and black-crowned night-heron *(Nycticorax nycticorax)* were the most utilized sentinels (Table 1). It was interesting that mallard, a widely distributed species, was studied in the USA and Spain. However, another globally distributed species, the night-heron, was only studied in the USA. White stork was biomonitored in 11 studies, as it proved to be a good sentinel species in the surrounding ecosystems of the Aznalcóllar spillage (Table 1). However, the most widely used Order of birds were passerines, with 51 species. This is not strange given that it contains the highest number of species among bird Orders. Nevertheless, passerine birds have historically been used as sentinels in fewer studies than raptors, probably because of their lower position in the food chain and their relatively short lifespan (Sánchez-Virosta et al. 2015). Despite this, passerine species have successfully been used as sentinel species of environmental pollution (Beyer et al. 2013; Sánchez-Virosta et al. 2015). They are especially useful for identifying local pollution, as they usually forage along small ranges (Costa et al. 2011), and they are easily captured. In addition, they can be severely affected by soil ingestion caused by their feeding behavior

(Hansen et al. 2011). The species more frequently biomonitored in metalliferous mining sites were American robin (*Turdus migratorius*) and tree swallow (*Tachycineta bicolor*) (Table 1). Just like black kites and white storks, the same species were used repeatedly in a particular mining site and by the same research group. This could give a false idea about how frequent is the use of these species as sentinels, but it also highlights that they are good sentinel species and have valuable data as a base for future studies. Also, there were studies that included several different passerine species, as they focused more on their diet and trophic position (granivores, insectivores, etc.) than on particular species (Abeysinghe et al. 2017; He et al. 2020). As stated before, birds are good as sentinel species because of several factors including the huge number of bird species worldwide, their ability to adapt to and thrive in disturbed habitats or their sensitivity to environmental pollutants (Smith et al. 2007). Moreover, non-destructive samples as eggs, blood, feathers or even feces have proved to be good indicators of metal exposure, and populations are relatively easy to quantify and monitor (Gómez-Ramírez et al. 2014). However, one important factor to take into account is their migratory behavior, because we need species capable to detect local sources of pollution at particular mining sites. Because of that, it is common to use nestling birds instead of (or in addition to) adults.

Mammal species were also widely used as sentinels in metalliferous mining sites. 100 studies included some mammal species, comprising 65 different species (Table 1). As in birds, oral consumption is the main pathway of contaminant exposure of wild mammals, either via food and water or through incidental oral exposure by such activities as grooming or soil licking (Smith et al. 2007; Beyer et al. 2007). Among them, 46 species were small mammals, including shrews, lagomorphs, and rodents. Small mammals are considered useful, sensitive, and accessible sentinels to evaluate ecological and health risks from pollutants (O'Brien et al. 1993). They can accumulate contaminants directly from soils, water, preys, and vegetation (Amuno et al. 2016) and they are considered to represent an intermediate stage between low and high trophic levels, although shrews can be considered predators with a high position in the food chain (Sánchez-Chardi et al. 2008). Most studied species were Algerian mouse (*Mus spretus*) (13) and wood mouse (*Apodemus sylvaticus*) (12), followed by deer mouse (*Peromyscus maniculatus*) (8) and bank vole (*Clethrionomys glareolus*) (7). Algerian mouse, wood mouse, and bank vole were only used as sentinel species in Europe, and deer mouse in North America, which are within their geographic range. Such small mammal species, and other less frequently studied, proved to be good sentinels of mining-polluted sites, usually showing higher metal body burden and more subclinical effects in contaminated than reference sites. However, not all small mammals showed similar patterns of metal accumulation and effects and it is essential to consider species-specific traits. Some important traits are spatial habitat use and preferences, diet preferences, differences in metabolism, and amount of incidental ingestion of soil particles during grooming activities or foraging of roots or soil invertebrates (van den Brink et al. 2011; Drouhot et al. 2014). Mammals other than small mammals were less represented in these studies (37), including 19 species. The major advantage of larger mammal over small mammal species is that they have the highest potential accumulation among all taxa of environmental contaminants (Hermoso de Mendoza García et al. 2008). In addition, large mammals have a relatively long lifespan, accumulating environmental effects over the years, and their shared physiological characteristics with human provide a valuable approach to compare mechanisms affecting human health (Hermoso de Mendoza García et al. 2008). A disadvantage could be the greater costs and difficulties of sampling compared to small mammals. Most studies used game species (14) or livestock species (9) foraging in contaminated areas as sentinel species, including 11 species of Artiodactyla and 1 species of Perissodactyla. Species such as red deer (*Cervus elaphus*) and wild boar (*Sus scrofa*) were included in 12 and 8 risk assessments, respectively. Both were mainly studied in two severely polluted derelict mine districts in Spain (Table 1, Almadén Mining District and Sierra Madrona and Alcudia Valley), but also in one study in Slovenia (Doganoc and Gačnik 1995). Regarding livestock species, cattle, sheep, and goat inhabiting near mine sites and foraging presumably contaminated food were used as sentinel species. Although some studies were carried out in Spain or the United Kingdom (Smith et al. 2010; Rodríguez-Estival et al. 2012), livestock species were more frequently used in developing countries such as Ghana (Bortey-Sam et al. 2015), Nigeria (Jubril et al. 2017, 2019), and Zambia (Yabe et al. 2011; Ikenaka et al. 2012) as sentinel species of both humans and wildlife health. Also, a wildlife species as the Kafue lechwe (*Kobus leche kafuensis*), an endangered species according to IUCN, was used to evaluate metal exposure and risks to wildlife and also to livestock and humans, as authors stated that lechwe shared grazing and drinking spots with cattle (Syakalima et al. 2001b). Despite consumption of livestock by humans is evident and therefore most developed countries have established maximum levels of trace elements for farm reared meat, there is a lack of legislation regarding game species meat (Taggart et al. 2011). An exception among game and livestock species of biomonitored wild hoofed mammal was hippopotamus (*Hippopotamus amphibious*) in Zambia (Mwase et al. 2002). Probably due to its conservation status, limited data is available on hippopotami, so this study evaluating the possible impact of trace element contamination from mining on hippopotami is very valuable. Also particularly interesting were the studies in the mammalian order Carnivora (6), including eight species. Wild carnivores are exposed to chemical contaminants through biomagnification, as they are usually at the top of the food chain, which makes them good sentinel species (Rodríguez-Estival and Mateo 2019). Black bear (*Ursus americanus*) was included in a study that examined adverse effects at all levels of biological organization in the USA (Peplow and Edmonds 2005). In Spain, the endangered Iberian lynx (*Lynx pardinus*) and other sympatric wild carnivores were assessed for trace elements concentrations after the Aznalcóllar spillage (Millán et al. 2008). They aimed to determine if selected trace elements posed a threat for the conservation of the Iberian lynx. This kind of studies involving endangered species are highly valuable, although usually these species cannot and should not be sampled due to their conservation status. As an alternative, already dead individuals could be used for the assessment (Millán et al. 2008), but better options are the use of non-destructive techniques and the use of surrogate sentinel species (Rodríguez-Estival and Mateo 2019).

Finally, reptiles were assessed in 11 studies, including nine different species of the order Crocodilia, Squamata, and Testudines, while only five studies included six amphibian species (Table 1). Although in the last few years scientific interest in the ecotoxicology of metals in reptiles has increased (Grillitsch and Schiesari 2010), this taxa continue to be underrepresented in ecotoxicological studies in general (Hopkins 2000; Linder et al. 2010), and in mining studies in particular. Effects of contaminants in amphibians have been more frequently evaluated than in reptiles, although tissue residues of contaminants are more common in reptiles (Linder et al. 2010). This underrepresentation of amphibians and reptiles in metalliferous mining risk assessments should be a matter of concern, given the great current declines and problems that both group of taxa are facing (Linder et al. 2010). Exposure pathways to contaminants are primarily through ingestion, dermal contact, and inhalation for both taxa, like other vertebrates. However, while ingestion is the main exposure route for reptiles, amphibians are the taxonomic group with higher potential for exposure through all routes. In addition, the contribution of the different routes changes through their life, due to their complex life cycle (Smith et al. 2007). Despite being the least studied group, as previously stated, reptiles have proved to be useful in risk assessment studies in mine sites. In particular, Moorish wall gecko (*Tarentola mauritanica*) proved to be useful for studying the bioavailability of some hazards metals after mining spills (Fletcher et al. 2006). Also, Mediterranean pond turtle (*Mauremys leprosa*) was useful revealing the bioavailability of metals in an area influenced by former mining (Martínez-López et al. 2017), and as sentinel of chronic metal pollution in two former mining areas (Ortiz-Santaliestra et al. 2019). Regarding the anuran species assessed, none of the species was used in more than one study (Table 1), and all of them only reported metal levels in different tissues, but none included biomarkers of effects. In addition, while in reptiles some of the studies tried to establish the usefulness of non-invasive and non-lethal tools, in amphibians all samplings were with lethal techniques. The progress in the use of non-invasive and non-lethal sampling techniques for exposure and effects should be essential, especially in taxa as herpetofauna with severe declining populations (Hopkins et al. 2001; Smith et al. 2007).

In summary, it is very important to take into account that the degree of accumulation or effect of metals on vertebrates is usually distinct for different species, depending on species-specific traits such as their trophic level, foraging behavior, and life history. In fact, some authors consider that "species" could be the best explanatory factor of the variation found in metal concentrations (Hernández et al. 1999). Some key characteristics for an animal to be considered a sentinel species are having a widespread distribution, a high trophic status, ability to bioaccumulate pollutants, having a restricted home range, being sensitive to pollutants, having a well-known biology, and that it could be maintained and studied in captivity as well as captured in sufficient numbers (Basu et al. 2007). However, species satisfying all these requirements are difficult to find, and both the confirmation of already established wild sentinel species and the search of new sentinel organisms that may serve as model organisms to infer and predict effects in other levels of biological organization become very important (Tovar-Sánchez et al. 2012). Finally, it is also important to highlight that, generally, the selection of the organism(s) of study is going to be site specific – that is, dependent on the species inhabiting the polluted site – and that it is important to consider all trophic levels, and to not completely rely on organisms at the top of the food chain (Peakall and Walker 1994).

3.4 Biomarkers

One of the main problems of ecotoxicological studies is the difficulty of giving an absolute definition of a body concentration range of metals that reflects "normal" conditions, given the variability among different species (Sánchez-Chardi et al. 2007c) and even within the same species. Usually, biomarkers of exposure are not enough to predict potential adverse effects, and therefore the use of biomarkers of effect is highly recommended (Mussali-Galante et al. 2013b). Although there are many biomarkers for measuring the effects of contamination on wildlife, a current challenge for ecotoxicologists is to identify biomarkers sensitive enough for the best possible diagnostic, allowing us to monitoring the impact of anthropogenic activities in each particular case (Sánchez-Chardi et al. 2007c). In this section, biomarkers of exposure and effect will be summarized. As a starting point, we considered a *biomarker of exposure* the detection of metal(s) in the different matrices, whereas a *biomarker of effect* included the response of organisms to these metals and their accumulation. Thus, biomarkers of effects included the alteration of any parameter associated with contaminant exposure, from biochemical parameters to population responses. Although we are going to summarize each biomarker separately, the order will be from the lowest level (macromolecular responses) to population level, including intermediate levels as biochemical, physiological, and individual level. Among studies developed in metalliferous mining sites, we found 79 publications which considered only biomarkers of exposure, measuring levels of different trace elements in a wide range of matrices. The remaining studies (107) included both concentrations of metals in some biological matrix and at least one biomarker of effect, although there were some exceptions which only assessed the biomarker of effect without exposure data.

3.4.1 Biomarkers of Exposure

Concentrations of numerous elements were reported in the different studies, including Ag, aluminum (Al), As, boron (B), barium (Ba), beryllium (Be), bismuth (Bi), calcium (Ca), Cd, cobalt (Co), chromium (Cr), cesium (Cs), Cu, Fe, gallium (Ga), Hg, potassium (K), lithium (Li), magnesium (Mg), Mn, molybdenum (Mo), sodium (Na), nickel (Ni), phosphorus (P), Pb, rubidium (Rb), sulfur (S), antimony (Sb), selenium (Se), silicon (Si), Sn, strontium (Sr), titanium (Ti), thallium (Tl), vanadium

(V), tungsten (W), and Zn. While some of them as Bi, K, Na, S, Si, and W were only assessed in one study (McIntyre and Whiting 2012), the elements most analyzed were As, Cd, Cu, Pb, Hg, and Zn. The two main reasons are that they are among the most extracted metals in mining areas (Table 1), and also that As, Cd, Pb, and Hg are among the most hazardous metals (Nordberg et al. 2007; ATSDR 2019). Levels of trace elements reported were highly variable, assessed in different compartments and expressed in different measures or different summary statistics. For this reason, it was almost impossible to summarize the different levels found for different species, taxa, or even compartments reported, so we included the levels found in the contaminated areas of the different studies in Table S1 (Supplementary Material) for references in future studies. Most studied tissues were liver and kidneys, given that they are the main organs of metal accumulation, although it depends on the toxicodynamics of each metal and on the timing, type, and intensity of exposure (Burger 2008; Sánchez-Virosta et al. 2015). In birds, a high number of studies also include blood levels, but in mammals and reptile species the blood was not so frequently included. Amphibians and reptiles studies frequently assessed whole body concentrations or carcasses, but also included internal organ burdens and muscle. Other tissues also studied were brain, testis, lung, heart, spleen, and bone. It is important to highlight the use of some non-destructive sampling apart from blood, such as hair in mammals, feathers and eggs in birds, or tail clips, scutes, and carapace in the case of reptiles. And even more important, the use of some non-invasive sampling, such as feces or regurgitated pellets, which have proved to be useful for the assessment of exposure to different metals (Chaousis et al. 2018). Besides animal matrices, some studies also included levels of metals in environmental compartments, such as soil, water, or plants, which is useful to assess the potential exposure in their environment.

Biomarkers of exposure are useful to assess if the sites are polluted and if this pollution is reaching the biota and the food chain. Almost all studies used a control area to compare levels found in animals sampled in the mining areas with those of the control area. This control area was usually a site with similar ecological characteristics to the polluted area, but without a known source of pollution. Usually, concentrations of metals found in the different matrices were higher in mining-impacted areas than control sites. However, unless evident clinical effects were shown, studies could only hypothesize the real damage suffered by the organisms by comparing the metal levels found with previous published data on sublethal effects in the same or related species. The problem is that studies usually used threshold levels established in previous laboratory studies which do not reflect the variable conditions in natural ecosystems that could also affect health status of organisms (Damek-Poprawa and Sawicka-Kapusta 2004). Sometimes they used levels established in field studies but for other species, without taking into account the variability and different sensibilities even between closely related species (Erry et al. 1999b). As stated at the beginning of the section, all this represents a problem and makes it difficult, if not impossible, to give an absolute definition of a body concentration range of metals that reflects "normal" condition (Sánchez-Chardi et al. 2007c).

3.4.2 Biomarkers of Effect

In the 107 studies found assessing the response of wildlife to the metalliferous mining pollution, a wide range of biomarkers of effect were used. Most studies included not only the effect, but also the measurement of metal concentrations associated with the effect. Furthermore, different biomarkers of effect were often included in a single study, including the responses at the same level of biological organization (e.g., cellular level) or different levels (e.g., cellular level and population level). Biomarkers were categorized according to the different physiological effects that they assessed.

Biochemical and Hematological Biomarkers

Metallothioneins. Metallothionein (MT) is a frequently used biomarker in terrestrial and aquatic animals, although only three studies in mining areas used this biomarker in cattle, Algerian mouse, and wild black rats (*Rattus sp.*). MT is a metal-binding low molecular protein which participates in detoxification of toxic metals, homeostatic regulation of essential metals, and protection against oxidative injury (Nakayama et al. 2013). As these proteins are induced by metals such as Cu, Cd, and Zn, an increase in MT levels or in their mRNA expression could reflect their accumulation (Laurinolli and Bendell-Young 1996; Nakayama et al. 2013). In Zambia, cattle blood showed higher Pb levels in the mining area than control, which was related to the induction of mRNA expression of one MT isoform (MT-2) in white blood cells, as well as various cytokines, reflecting effects on the immune system (Ikenaka et al. 2012). Also in Zambia, Cu (range 1.85–20.85 µg/g ww in liver; 1.14–5.8 µg/g ww in kidney) and especially Zn (18.3–60.3 µg/g ww in liver; 10–25.7 µg/g ww in kidney) levels showed a positive effect on MT mRNA expression in liver and kidney of wild back rats (Nakayama et al. 2013). However, Pb and Cd levels were low from toxicological points of view, and Cd concentrations in kidney showed a negative effect on the MT-2 mRNA expression levels. MT levels were also used as a biomarker for the first time in the Algerian mouse in an abandoned mining area in Portugal. Mice from the mining area showed elevated Fe, Se, and MTs levels in liver in comparison with the reference area. In addition, MTs levels were positively correlated with liver elemental contents, particularly Cu in autumn (4.1±0.3 µg/g ww) and Fe in winter (169±20 µg/g ww). For comparisons, conversions from dry weight (dw) to wet weight (ww) were done using same moisture levels as (Ma 2011). In birds, no study in mining areas used this biomarker and little is known about MT concentrations in free-living birds in field conditions, especially in terrestrial bird species (Vanparys et al. 2008). These authors affirmed that their results in a metallurgic smelter area support the applicability of MT in birds studies and highlight the importance of the relation between MT and Cd as has been previously shown in several species (Vanparys et al. 2008). Although this biomarker could show sublethal biochemical alterations and therefore be useful as an early warning,

MT expression levels are not specific to metal pollution (Nakayama et al. 2013). Thus, in a context of metal mixtures, it is important to consider the possible metal–metal interactions, given that it could be possible that Cu exposure of mice may have stimulated the induction of MT and this increase, in turn, possibly enhanced the uptake of Zn and Cd (Laurinolli and Bendell-Young 1996).

ALAD activity. Among biochemical parameters, the activity of the enzyme delta-aminolevulinic acid dehydratase (ALAD) is considered to be the most sensitive effect biomarker for Pb exposure (Rodríguez-Estival et al. 2012). Because of that, it has been widely used in studies of wildlife inhabiting contaminated sites (Hansen et al. 2011), and 19 of the reviewed papers measured this parameter (18% of papers that included some biomarker of effect). This enzyme activity, which catalyzes the formation of a precursor of heme, is inhibited in animals when they are exposed to Pb. Although it is mainly used as an indication of Pb exposure, high levels of inhibition are considered to be indicative of injury. In fact, regulations in the USA established this threshold in >50% inhibition (NRDA regulation 43 CFR 11.62, 2004; Johnson et al. 1999). However, other studies in birds showed that it depends on several factors, such as age or species. For example, adult birds or precocial hatchlings would need a continued 80% or higher blood ALAD inhibition for decreased hemoglobin, but only 45–59% inhibition is enough in altricial nestlings (Beyer et al. 1988; Gómez-Ramírez et al. 2011). Some direct consequences of the enzyme inhibition are the accumulation of ALA, which causes neuropathogenic effects and stimulates reactive oxygen species production, and the decline in hemoglobin and hematocrit, causing anemia (Gómez-Ramírez et al. 2011). In the assessment of ALAD inhibition, activity levels were usually compared with animals from an uncontaminated area that theoretically have basal activity, or sometimes were compared with levels previously reported for the same or close species. One of the main problems is that Pb levels at which ALAD is inhibited could be very different depending on the scenario, and enzymatic inhibition induced by Pb can differ between species, age (inhibition is generally higher in adults than nestlings), and even physiological status (Gómez-Ramírez et al. 2011). For comparisons, conversions of Pb concentrations in blood from mg/kg to µg/dl were done using blood density of 1 g/ml, and from dry weight (dw) to wet weight (ww) were done using same moisture levels as Franson and Pain (2011).

In waterfowl, ALAD inhibition was usually very high in different species, including tundra swan, mallard, wood duck (*Aix sponsa*), Canada goose, common pintail (*Anas acuta*), green-winged teal (*Anas crecca*), ring-necked duck (*Aythya collaris*), and lesser scaup *(Aythya affinis)*. Among waterfowl species, all studies (7) included mean inhibition levels ≥50%, the legal level indicative of Pb poisoning injury in the USA. In fact, all exposed Canadian geese showed ≥50% inhibition. Furthermore, most inhibition levels were higher than this threshold. Tundra swans found sick and moribund showed almost complete inhibition (98%) of ALAD with Pb levels in blood as high as 330 µg/dl (Blus et al. 1999). Beyer et al. (2004) also reported one female mallard with 98% ALAD inhibition that showed 128 µg/dl Pb concentration. Adult waterfowl mean ALAD inhibition ranged from 59 to 97% at mean Pb levels between 20 and 86 µg/dl and Pb geometric mean between 40.9 and

180 μg/dl. Young birds showed 65–87% ALAD inhibition at geometric mean concentration of Pb between 28.3 and 37 μg/dl (Blus et al. 1991, 1993, 1995, 1999; Henny et al. 2000; Beyer et al. 2004; van der Merwe et al. 2011). Passerines were also examined, including species such as American dipper (*Cinclus mexicanus*), American robin, Northern Cardinal (*Cardinalis cardinalis*), Brown thrasher (*Toxostoma rufum*), Pied flycatchers (*Ficedula hypoleuca*), song sparrows (*Melospiza melodia*), Swainson's trushes (*Catharus ustulatus*), and tree swallows. All studies (5) assessing ALAD inhibition in adult passerines showed ≥50% mean inhibition levels, with more than 50% of birds above this threshold in most of them. However, in studies including nestling birds (3), although some of the nestlings showed ≥50% ALAD inhibition, mean inhibition was lower. Mean ALAD inhibition in adult passerines ranged from 57 to 90% corresponding to mean Pb concentrations in blood between 30 and 135 μg/dl and a median Pb level of 15.1 μg/dl in one species (Strom et al. 2002; Beyer et al. 2004, 2013; Hansen et al. 2011). Also, at a mean Pb level of 1.93 μg/g ww in liver, inhibition reached 51% in adult song sparrows (Johnson et al. 1999). In addition, some birds with their highest Pb levels between 48 and 87 μg/dl in blood showed 91–95% ALAD activity inhibition (Beyer et al. 2004). Nestlings from different species showed mean ALAD inhibition between 26 and 43%, and pied flycatcher nestlings averaged 41 μg/dl blood Pb in nestlings with ≥50% ALAD inhibition (Custer et al. 2002a; Strom et al. 2002; Berglund et al. 2010). In some cases ALAD inhibition reached levels (81–93%) associated with 50% mortality in songbirds (Beyer et al. 1988). Raptor species were also evaluated for Pb exposure and effects through ALAD activity. Species studied included Eurasian Eagle Owl, Osprey (*Pandion haliaetus*), American kestrel (*Falco sparverius*), and Northern Harrier (*Circus cyaneus*). Adult birds of prey (2 studies) showed ≥50% ALAD inhibition, with inhibition levels between 64 and 81% in individuals with geometric mean Pb blood levels of 20–46 μg/dl (Henny et al. 1991, 1994). Regarding nestlings, some species (3) showed mean ALAD inhibition above the threshold (52–60%) with geometric mean concentrations of Pb in blood ranging 9–24 μg/dl (Henny et al. 1991, 1994; Gómez-Ramírez et al. 2011). However, two studies showed 35% ALAD reduction at a geometric mean of 6.7 μg/dl Pb in blood (Henny et al. 1994), and mean 28% inhibition at a mean Pb blood level of 7.84 μg/dl (Espín et al. 2015). Nevertheless, individuals from this study showed 50% ALAD inhibition at levels >10 μg/dl and 79% inhibition at levels >20 μg/dl of Pb in blood.

The majority of studies (17) were in bird species, given that it has been suggested that ALAD inhibition could be more sensitive and its effects more severe in birds than mammals (Vanparys et al. 2008; Gómez-Ramírez et al. 2011). However, it has proved to be useful also in mammals species (Rodríguez-Estival et al. 2012), including three species in two different studies. Both studies showed mean ALAD inhibition above the threshold indicative of injury. White-footed mice (*Peromyscus leucopus*) showed a 68% reduction in ALAD activity at a mean blood Pb concentration of 20.7 μg/dl (Beyer et al. 2018). Cattle and sheep from mining areas had lower ALAD activity, cattle reduced by 68% (Pb blood geometric mean 14.74 μg/dl) and sheep by 73.8% (Pb blood geometric mean 2.25 μg/dl) (Rodríguez-Estival et al. 2012). These authors showed that ALAD activity was also inhibited in the mining area within the range currently considered as background levels in blood (<6 µg/dl) for livestock. In humans, the Agency for Toxic Substances and Disease Registry established that ≥50% ALAD inhibition was associated with a blood Pb concentration above 20 µg/dl (Beyer et al. 2018).

As shown by the summarized studies, the activity of ALAD enzyme proved to be a sensitive and accurate biomarker for Pb exposure and effect. Most sentinel species used for assessing metal pollution in different mining sites showed levels above the legal requirement for defining Pb poisoning injury (≥50% ALAD inhibition) according to the USA legislation. Sometimes, this level could be reached at tissue concentrations lower than the concentrations associated with visible signs of lead poisoning (van der Merwe et al. 2011). But some of the bird species analyzed in these studies were exposed to very high levels of Pb in the highly contaminated Coeur d'Alene River Basin (USA), and Pb levels were compatible with subclinical and clinical poisoning levels (Franson and Pain 2011). Besides ALAD activity, other related biomarkers of effect were used at the same time in some studies. For example, protoporphyrin, an intermediate in heme synthesis, which is less sensitive than ALAD, given that it increases at higher Pb concentrations than when ALAD is inhibited (Henny et al. 1994). Hematological parameters, such as hematocrit and hemoglobin levels, were also employed in some of the ALAD inhibition studies, but again are less sensitive markers that were not significantly affected in many cases. In some studies, hematocrit and hemoglobin were significantly depressed only in moribund swans (Blus et al. 1999). In contrast, some sentinel species showed increased levels of hematocrit in mining sites. According to the authors, this may be due to a compensatory response of the organisms associated with the decrease in ALAD activity, increasing hematocrit values to improve the oxygen-transport capacity of the blood (Rodríguez-Estival et al. 2012; Espín et al. 2015).

Blood-related parameters and immune system. These two parameters were included in the same subsection because almost all studies that included blood cell counts included white blood cells, which are directly related to the immune system. 25 studies assessed effects in blood molecules/cells or immune system. Serum chemistry values and cell blood counts were assessed in two lemur species (*Avahi laniger* and *Lepilemur mustelinus*) inhabiting a protected forest near a cobalt mine that recently began ore extraction. Levels found were similar to base levels of other lemur species, and authors stated that levels found allowed them to established baseline levels for future monitoring in the same area (Junge et al. 2017). In Portugal, shrews showed higher hepatic Cd levels in a mining area than control but did not show significant changes on hematological parameters (Marques et al. 2007). In another Portuguese mine, Algerian mice were assessed for a wide range of hematological parameters, but only hemoglobin concentration was higher in the polluted site (Nunes et al. 2001b). However, we must be cautious when measuring only those parameters but not metal levels in tissues. Lidman et al. (2020) found that hemoglobin level in nestling pied flycatcher inhabiting close to an old mine was higher than those from control area. However, hemoglobin was not directly related to Pb accumulation and they found that other factors, as prey availability, had more influence on hemoglobin than Pb levels. Another studied blood-related parameter was porphyrin, but measured through non-invasive samples. As disruption of heme synthesis by metal exposure generates a surplus in the production of different heme precursors, some of them are excreted at higher levels through urine or feces (Martinez-Haro et al. 2013). Thus, waterfowl species studied after Aznalcóllar spillage showed alterations in coproporphyrin and biliverdin ratios in feces, which were related to Pb and As concentrations in their tissues (Mateo et al. 2006; Martinez-Haro et al. 2013). Furthermore, immune response could also be altered by metal pollution. In fact, the immune system is recognized as a target of Pb-induced toxicity (Rodríguez-Estival et al. 2013b). These potential effects were studied in song sparrows, which showed significantly increased mean white blood cells. Although overall health of birds appeared normal, this increase could be a possible low-level immune response, also related with higher mean levels of Hg in blood and Pb in feathers (Lester and van Riper 2014). Hg was also apparently affecting immune function of western and Clark's grebes (*Aechmophorus occidentalis* and *Aechmophorus clarkia*) (Elbert and Anderson 1998). Those grebes showed levels of Hg in kidney positively correlated to percent blood heterophils and negatively correlated to percent eosinophils (white blood cells). Also in young snowy egrets and black-crowned night-herons Hg-associated effects to the immune system included alterations in white blood cells (Hoffman et al. 2009). In addition, positive correlations between total Hg and plasma uric acid and phosphorus were found, suggestive of renal stress. In this study drought may have exacerbated these Hg-related effects, highlighting the importance of environmental factors. Another bird assessed was white stork, in which cell-mediated immune response (CMI) was negatively related to Cu levels (only significant for one year). However, there was no evidence that metal levels had an effect on CMI (Baos et al. 2006b). Red deer and wild boar were assessed for an association between Pb exposure and the transcription profile of some cytokines (Rodríguez-Estival et al. 2013b). Authors concluded that Pb may alter the immune response, although no evidence of direct relationship was found. Finally, the mRNA expression of various cytokines in white blood cells was measured in cattle, which were increased at the same time as Pb and Cd blood levels (Ikenaka et al. 2012).

Oxidative stress biomarkers. Other commonly used biomarkers of effect are those related with oxidative damage. Mining related effects were assessed through parameters related to oxidative stress in 26 studies (24% of papers that included some biomarker of effect). Oxidative stress is produced in organisms when there is an imbalance between the production of reactive oxygen species and the antioxidant capacity of the organism. Metals are known to increase the natural formation of reactive oxygen species (ROS) that, in excess, can react with biological molecules and induce lipid, protein, and DNA oxidation, causing adverse health effects (Berglund et al. 2007; Nakayama et al. 2013). To protect against these effects, organisms depend on their antioxidant system, which is integrated by compounds produced endogenously and dietary antioxidants (Rodríguez-Estival et al. 2011a). Among the endogenous compounds is glutathione (GSH), a low molecular weight antioxidant, which is the main compound responsible for overall cellular redox regulations. Other endogenous compounds are enzymes such as superoxide dismutase (SOD), catalase (CAT), or glutathione reductase (GR), with different mechanisms of action. The second group of antioxidants includes vitamins and nutritional trace minerals that can reduce oxidative stress, such as vitamin A (retinyl esters and retinols) and E (α-tocopherol) (Rodríguez-Estival et al. 2011a). Thus, levels of these compounds or levels of activity of the enzymes have been widely used for assessing metals adverse effects on animals. Also, direct damage produced by oxidative stress can be measured through the by-products of lipid peroxidation, as thiobarbituric acid reactive substances (TBARS) or malondialdehyde (MDA).

The majority of studies assessed oxidative damage in mammals (19 studies). Algerian mice inhabiting areas affected by the Aznalcóllar mine spill showed elevated glutathione-*S*-transferase (GST) activity and altered activity of antioxidant enzymes such as glutathione peroxidase (GPx) and glucose-6-phosphate dehydrogenase (G6PDH) six months after the accident (Ruiz-Laguna et al. 2001). One and two years after the spillage, effects were still noticeable, as mice showed high levels of Pb, Cd, and As that were significantly correlated with CAT, GR, and G6PDH activities, but also with levels of MDA and oxidized glutathione (GSSG) (Bonilla-Valverde et al. 2004). The same species, inhabiting an active mine in Portugal, showed elevated levels of Mn, Fe, and Se in liver and higher GST activity than control site, as well as slightly lower SOD activity, indicating biochemical stress (Viegas-Crespo et al. 2003). On the contrary, greater white-toothed shrews (*Crocidura russula*) from a derelict mine showed a reduced GST activity and significant correlations between Pb, Cu, Mn, and Cr in liver and GPx activity (Sánchez-Chardi et al. 2008). Shrews also showed higher Cd and Ni levels in liver in an old mining-polluted site, but without significant changes on redox balance in this case (Marques et al. 2007). Other small mammal species, the meadow vole (*Microtus pennsylvanicus*), had elevated As levels in liver related to diminished GSH levels (Saunders et al. 2009, 2010). A different biomarker was used in a study on wild *Rattus* spp. which used the gene expression of heme oxygenasa-1 (HO-1), a cellular stress protein, instead of using measures of antioxidant molecules or enzymes, resulting a good indicator for Cr exposure (Nakayama et al. 2013). White-footed mice from Southeast Missouri Lead Mining District also showed lower levels of biomarker of oxidative stress (GSH, thiols, and protein-bound thiols) at contaminated than reference site. However, this oxidative stress caused by Pb was not severe enough to affect lipid peroxidation (Beyer et al. 2018). A lagomorph species, the wild snowshoe hares (*Lepus americanus*), showed increased levels of lipid peroxidation and reduced SOD, CAT, GPx, and GSH as effects of chronic As and Cd exposure in a mine site (Amuno et al. 2018b). Oxidative stress was also obvious in balochistan gerbils (*Gerbillus nanus*) which showed elevated levels of diverse metals (Cd, Pb, Hg, Zn, Cu, Fe, As, and V in kidney, liver, and lungs) and increased lipid peroxidation (MDA) and nitric oxide (NO) as well as decreased GSH, CAT, and SOD in liver and kidney (Almalki et al. 2019b).

Game mammal species, such as wild boar and red deer, have also been widely assessed for oxidative damage in hunting estates located in old Pb and Hg mining districts in Spain. Metal pollution effects on sperm quality of red deer were related to low Cu and Se levels in testis, and associated low SOD and GPx activities in testis and spermatozoa (Reglero et al. 2009a). Changes in the fatty acid profiles of spermatozoa were also associated with low Cu levels (Castellanos et al. 2010), and high Pb levels in testis were related to lower acrosome integrity and higher GPx activity. In addition, damage in chromatin condensation could have been triggered by oxidative stress, because the high GPx activity found on deer may reflect an attempt to maintain the redox balance in spermatozoa (Castellanos et al. 2015). Apart from sperm quality, levels of GSH and fatty acid in deer livers were also reduced at the same time as elevated Pb, Cu, Cd, and Se levels were found (Reglero et al. 2009b). Hg effects were also evaluated in deer, which showed higher Hg levels in kidney and liver in individuals living in areas closer to the mine site (Berzas Nevado et al. 2012). These Hg concentrations and Hg:Se molar ratio negatively correlated with GSSG. According to the authors, this reduction of GSSG could be a compensatory effect by the organism in response to early signs of oxidative stress. Various studies were developed both in red deer and wild boar in a Pb-contaminated area. Red deer showed high concentrations of Pb in bone and liver related to lower GSH, α-tocopherol, and retinyl esters as well as higher free retinol and TBARS. However, wild boar showed lower retinyl esters and increased free retinol, GSH and α-tocopherol. Also, GPx and Se were higher in boar than deer, which may indicate that red deer is more sensitive than boar to oxidative damage (Rodríguez-Estival et al. 2011a). Vitamin A and E status of these species was also altered. Deer showed lower retinyl esters in liver and lower free retinol, α-tocopherol, and a retinyl ester in testis, which overall indicated a higher internal usage of these antioxidants. Also, wild boar showed lower retinyl ester but higher free retinol in liver, which may indicate that vitamin A was mobilized to cope with induced oxidative stress. Moreover, a significant relationship was found between liver α-tocopherol and bone Pb in boar, which could indicate long-term effects on antioxidant levels (Rodríguez-Estival et al. 2011b). Also in this area, both species showed high Pb levels in spleen, liver, and bone associated with a depletion of spleen GSH levels and disrupted activity of antioxidant enzymes GPx and SOD (Rodríguez-Estival et al. 2013b).

Six studies regarding oxidative stress biomarkers were carried out in bird species. Eurasian eagle owl individuals showed variations in GST and CAT activities compared with those from non-mining areas due to elevated levels of Pb and Hg. However, the lack of difference in oxidative damage to membrane lipids (TBARS) among areas seems to indicate that their antioxidant system was able to deal with the oxidant species (Espín et al. 2014a). On the contrary, young snowy egrets inhabiting a highly Hg contaminated area showed elevated lipid peroxidation as well as decreased hepatic total thiol concentration and GR activity. However, these birds also showed decreased GSSG, which was probably a compensatory mechanism in response to oxidative stress (Hoffman et al. 2009). Young double-crested cormorants containing high Hg concentrations also showed increased lipid peroxidation, low activities of glutathione related enzymes, and an increase in the ratio of GSSG to reduced GSH (Henny et al. 2002). Laughing doves (*Spilopelia senegalensis*) from Saudi Arabia also showed increased lipid peroxidation and NO as well as reduced GSH, SOD, and CAT than those of reference site. These doves showed increased levels of multiple metals (Pb, Cd, Hg, V, As, Cu, Zn, and Fe) in kidney, liver, and lung (Almalki et al. 2019a). Shaoxing duck (reared under natural conditions near mining areas) were evaluated for oxidative damage, as a risk assessment to human health. Ducks showed higher levels of Hg and Se in samples from the contaminated area as well as higher levels of GSH and activities of SOD and GPx. Authors explain these increases as an adaptive response of the redox-defense system, probably caused by the protective role of Se (Ji et al. 2006). Finally, a different component of the antioxidant system, plasma carotenoids, was analyzed in black-necked grebes (*Podiceps nigricollis*). Grebes were assessed for metal pollution during their moulting period, showing high blood levels of As, Hg, and Zn in comparison with toxicity thresholds (Rodríguez-Estival et al. 2019). Plasma carotenoids showed a positive association with As, and eye redness (also dependent on carotenoids) was negatively affected by As levels. Grebes were possibly undergoing an allocation trade-off between using available carotenoids for maintaining cellular redox balance (potentially disturbed by metal exposure) and for carotenoid-based signaling. Finally, one study was developed in a reptile species, the Mediterranean pond turtle. Turtles showed mean higher Pb blood levels in a Pb mining district and higher Hg levels in sites close to former Hg mines. Individuals from the Pb-contaminated area showed lipid peroxidation and also high levels of SOD, which apparently were insufficient to counteract action of ROS. Turtles from the Hg contaminated area showed increased activity on GPx and reduced level of GSH (Ortiz-Santaliestra et al. 2019).

Summarizing, although not all studies showed direct damage effect through lipid peroxidation measurements, most studies showed oxidative stress related effects caused by metal pollution. The most analyzed parameters were the levels of GSH (and GSSG) and the activity of related enzymes such as SOD, GPx, and CAT. Lipid peroxidation measured though TBARS or MDA was also frequently used in the studies. However, dietary antioxidants were the least used parameters. This is in agreement with Isaksson (2010), which concluded in her meta-analysis that GSH generally show the strongest response to pollution, together with its associated enzymes GST and GPx. In addition, we must be cautious in oxidative damage assessments, since animals may physiologically compensate for moderate exposure by increasing their antioxidant defenses (Beyer et al. 2018) as shown in some of the studies. Finally, the main disadvantage of these biomarkers is that they are usually measured using invasive sampling methods, mainly in liver and kidney. In fact, the quoted meta-analysis (Isaksson 2010) suggests that invasive sampling is more reliable than less-invasive sampling (as blood, urine, or sperm). However, it recommends the use of less-invasive sampling by using the most sensitive antioxidants (GSH and associated enzymes) along with a larger sample size to maximize power.

Genotoxic Parameters

Genotoxicity. Analysis of DNA alterations has shown to be a highly suitable method for evaluating environmental genotoxic pollution, allowing us to detect and quantify genotoxic damage without a detailed knowledge of the contaminants (Tovar-Sánchez et al. 2012). 19 of the reviewed studies assessed DNA damage through different techniques and in different species. Comet assay or alkaline single cell gel electrophoresis was one of the techniques used for evaluating DNA damage. This test has proved to be extremely sensitive and useful, showing clear dose–response relationships even at relatively low levels of naturally occurring genotoxicants (Baos et al. 2006c). Comet assay in peripheral blood lymphocytes was performed in white stork and black kite to assess genotoxic effects after the Aznalcóllar accident for 4 years after the spillage. Nestlings of both species born near the spillage area showed an increase of at least 2- to 10-fold in their genetic damage, in comparison with those individuals born in non-polluted areas (Pastor et al. 2001, 2004). Other study in the same area found correlations between DNA damage and blood levels of As in storks, and Cu and Cd in kites. However, relationships were only significant some years, and this inter-annual inconsistency does not allow to establish a cause–effect link. Because of that, they concluded that metals alone do not seem to completely explain the DNA damage found (Baos et al. 2006c). Some small mammal species also showed different levels of DNA damage assessed by comet assay. Algerian mice living in areas affected by Aznalcóllar spill showed high values in comet assay performed on peripheral blood leukocytes (Festa et al. 2003; Udroiu et al. 2008), showing also elevated levels of As and Zn in liver (Udroiu et al. 2008). Algerian mouse was also used as sentinel species in other mining-polluted area, showing elevated levels of genetic damage (Mateos et al. 2008). Other two species, plateau mice (*Peromyscus melanophrys*) and Southern pygmy mice (*Baiomys musculus*), living in abandoned mine tailings, showed significant DNA damage and elevated levels of Zn, Ni, Fe, and Mn in bone and liver. In addition, pygmy mice showed decreasing DNA damage with increasing distances from the tailings (Tovar-Sánchez et al. 2012). Also Nelson's pocket mice (*Chaetodipus nelson*) and Merriam's kangaroo rat (*Dipodomys merriami*) showed higher DNA damage in rodents living in mining areas, where they had elevated levels of As, Pb, and Cd in liver and kidney tissues (Jasso-Pineda et al. 2007). Kangaroo rat also showed elevated DNA damage in a contaminated site where its estimated oral As exposure was high (Espinosa-Reyes et al. 2010).

Another genotoxic technique is the micronucleus test (MN), an easy and quick method to obtain information about the clastogenic effect of environmental pollutants (Sánchez-Chardi and Nadal 2007). Micronuclei are small, extra-nuclear bodies resulting from chromosome breaks and/or whole chromosomes that did not reach the spindle poles during cell division (Kirsch-Volders et al. 2011). These particles are naturally occurring in all the organisms, but metals can increment their numbers, as showed in the following studies. Greater white-toothed shrews affected by mining pollution had higher frequencies of MN in blood erythrocytes than those individuals from the reference areas, having also high levels of Fe, Pb, Hg, Cd, Mo, and Ni in liver (Sánchez-Chardi et al. 2008) or only Cd and Hg in liver (Sánchez-Chardi et al. 2009). Meadow voles also showed elevated levels of micronucleated red blood cells associated with high concentrations of As in their tissues (Saunders et al. 2009,

2010). The frequency of total MN was also high in Algerian mice in areas affected by Aznalcóllar mining spillage, but DNA damage could also be due to the presence of agriculture pollutants in the environment (Tanzarella et al. 2001). In other study at the same area, this species also showed elevated MN frequency and high levels of As and Zn in liver (Udroiu et al. 2008).

A different technique used to determine potential DNA damage was flow cytometry, a rapid and cost-effective method used to detect chromosomal lesions (Hays and McBee 2007). It was used in blood and bone samples of various small mammals and birds species in the same study. They found no abnormalities in DNA damage, despite showing higher levels of Pb and Cd in blood and liver in the contaminated than in the reference area (Brumbaugh et al. 2010). Other study using flow cytometry in red-eared slider turtle (*Trachemys scripta*) did not find significant differences in nuclear damage between areas either, although a significant higher frequency of aneuploidy was found in turtles from the polluted area (Hays and McBee 2007). Finally, no evidence of chromosomal damage on flow cytometric analyses was found in tree swallows (Custer et al. 2002a).

Another two different biomarkers of DNA damage were used in two studies. Levels of 8-hydroxy-deoxy-guanosine (8-OH-dG) in liver of white-footed mice were used as a biomarker of DNA damage resulting from oxidative stress. Mean values of 8-OH-dG did not differ among contaminated and control site, so there was no perceptible damage to hepatic DNA (Beyer et al. 2018). Besides, DNA alterations examined in spermatozoa of red deer were present even at sublethal levels of Pb exposure. This affected sperm quality and could have potential consequences on the reproduction and the offspring (Castellanos et al. 2015).

The most used technique in order to assess genotoxic damage was the comet assay, which was carried out in nine different studies, including two bird species and five small mammal species. As stated before, this test has proved to be extremely sensitive and useful in previous studies, and results of reviewed risk assessments of metalliferous mining pollution confirmed that. DNA damage was elevated in most of the mining-polluted areas in several species. Regarding micronucleus test, six studies assessed DNA damage through it, all of them in small mammal species. However, this test has also proved to be useful in bird species (Barata et al. 2010). MN was also an accurate test that detected DNA damage in all studies reviewed. Comet assay and MN test are thought to be almost equally sensitive in assessing genotoxicity. However, while the comet assay detects primary repairable DNA damage, the MN assay detects more persistent DNA lesions or aneugenic effects that cannot be repaired (Klobučar et al. 2003). The third most used technique was flow cytometry, although this technique fails to detect chromosomal lesions in most reviewed studies, including various small mammals and bird species and one reptile species. An advantage of these techniques is that they can be performed in blood, a non-destructive sample. Finally, it is remarkable that most studies assessing genotoxic damage were carried out in Spain and Portugal, including various assessments developed after the Aznalcóllar spillage.

Health Status at Individual Level

Morphological parameters. Although morphological biomarkers are not usually indicative of a specific pollutant, they are sensitive for monitoring the impact of anthropogenic activities. Thus, they can provide useful data for the identification of deleterious consequences on populations inhabiting polluted environments (Sánchez-Chardi et al. 2007c). Some studies (15) evaluated the potential correlations between contaminants levels and different internal and external measurements. Parameters such as body weight and length, organs weight, asymmetry, and indices related to those parameters were used to assess body condition in different species. In greater white-toothed shrew, hepatic weight was higher in a contaminated area with increased levels of Fe, Pb, Hg, Cd, Mo, and Ni (Sánchez-Chardi et al. 2007b). Shrews also showed increased hepatic and renal weights and decreased body weight in another area related to increased liver and kidney Pb, Cd, and Hg levels (Sánchez-Chardi et al. 2009). Algerian mice inhabiting a mining area showed lower body weight and lower spleen and kidney masses than in the reference area (Nunes et al. 2001b). Also wild rats had lower body weight with increasing renal Pb levels (Nakayama et al. 2011). However, wild rats and Algerian mice from an abandoned mine area in Portugal did not show differences when compared to the reference site in body and organs weights, except a slight decrease of spleen weight, which could be explained by spleen atrophy (Pereira et al. 2006). A different parameter used in Algerian mice was the assessment of size and shape variations of the mouse mandible in two abandoned mines and a reference area. Results revealed differences in mandible shape which were suggestive for an effect of environmental quality on normal development pathways (Quina et al. 2019). In a remediated site, deer mice (*Peromyscus maniculatus*) showed lower body fat and mass as well as smaller body size (Coolon et al. 2010). However, white-footed deer mice had larger body mass in a contaminated site than in the reference area (Phelps and Mcbee 2010). These mice also showed poor dental condition and extensive mottling of the incisors, a reliable indicator of exposure to metal. Metalloid As also affected morphological parameters of various small mammal species which showed correlations between As levels and various organ masses and body measurements (Drouhot et al. 2014). In France, a risk assessment showed that past mining still has impact even in currently protected areas. Wood mice showed a negative correlation between kidney Pb concentration and Scaled Mass Index (SMI), an index often defined as a measure of the energetic state of the animal. However, metals did not exhibit any clear relationship with relative size of internal organs nor fluctuating asymmetry (Camizuli et al. 2018). Fluctuating asymmetry was also assessed in dental characters of Algerian mice, which showed smaller tooth traits in the contaminated area and increased developmental instability (Nunes et al. 2001a). But not only small mammals were assessed for effects with these measurements. The red deer showed higher testis mass and size at the same time that Cu levels were diminished, which can cause negative effects on sperm quality (Reglero et al. 2009a). A reptile species, the giant sungazer lizard (*Smaug giganteus*), showed body measurements negatively correlated with contaminants levels, although not significantly (McIntyre and Whiting 2012). Some bird species that showed alterations in these parameters were the wood duck (*Aix sponsa*), whose body mass negatively correlated with high blood Pb levels (Blus et al. 1993). Also, moulted female black-necked grebes showed an increase in mass positively correlated to an increased exposure to As during two months. This was probably due to higher consumption rates of their main food item, the brine shrimp, which showed high loads of As (Rodríguez-Estival et al. 2019). Finally, song sparrows showed no significant site differences in body mass residuals despite showing differences in metal exposure (Lester and van Riper 2014).

Reviewed studies showed high variability among the different assessments. Generally, lower body mass and body sizes were related to elevated levels of different metals. But these parameters highly depend on other intrinsic and extrinsic factor as age, sex, food availability, or season. Because of that, this kind of measurements could be indicative of poor health condition, but the severity of the effects should be assessed in conjunction with additional biomarkers, such as genetic, histopathological, or reproductive parameters (Nunes et al. 2001b).

Histopathological parameters. Histopathological analyses have been used for the evaluation of internal damage at individual level. This data combined with levels of metals in tissues provides valuable information about the condition of living organisms (Damek-Poprawa and Sawicka-Kapusta 2004). Histopathological examination can be performed in different organs, such as liver, kidney, or pancreas, and it involves the death of the studied individual. Although it was usually used in trapped and euthanized animals, it was also used in the necropsies of diverse waterfowl for determining the cause of death. In the CDA river basin diverse waterfowl species showed histopathological lesions related with lead poisoning (Sileo et al. 2001). Also, a postmortem examination of a trumpeter swan revealed severe histopathological lesions in the pancreas along with very high levels of Zn (3,200 µg/g ww). Consequently, it was identified as Zn toxicosis (Carpenter et al. 2004). High Zn levels were also found in the exocrine pancreas of Canada geese and mallards which showed degenerative abnormalities in that organ (Sileo et al. 2003). Years later, Canada geese still showed histopathologic signs in pancreas of Zn poisoning associated with elevated pancreatic Zn concentrations (van der Merwe et al. 2011). However, greylag geese assessed after the Aznalcóllar spillage showed mononuclear infiltrates in liver and kidney, but did not show other more specific lesions of Pb or Zn poisoning (Mateo et al. 2006). In other study, diverse waterfowl species were examined without finding proximal tubule cell necrosis associated with Cd poisoning despite having levels (141 µg/g ww in kidneys) associated with nephropathy (Beyer et al. 2004). Hg effects through histopathological analysis were assessed in cormorants, night-herons, and egrets. Young individuals showed several histological damage in different tissues as liver, kidney, peripheral nerves, bursa, and thyroid, correlated with Hg concentrations (Henny et al. 2002; Hoffman et al. 2009). These biomarkers were also used for health assessment of some bird species. For example, American robins showed renal damage associated with Pb poisoning (Beyer et al. 2013). Species such as Ornate tinamou (*Nothoprocta ornate*) and Darwin's nothura (*Nothura darwinii*) showed elevated Cd levels and related thesaurismosis (storage disease) in proximal convoluted renal tubules (Garitano-Zavala et al. 2010). Also related with Cd toxicity, white-tailed ptarmigan (*Lagopus leucurus*) showed accumulation of this metal and related kidney damage (Larison et al. 2000). However, tree swallows, violet-green swallows (*Tachycineta thalassina*), and mountain chickadee (*Poecile gambeli*) showed some histological changes in kidney but they were non-significant (Custer et al. 2009). Gold mining in Saudi Arabia was probably causing multiple histological alterations in liver, kidney, and lung of laughing doves along with elevated levels of multiple metals (Pb, Cd, Hg, V, As, Cu, Zn, and Fe) (Almalki et al. 2019a). Besides birds, these types of analysis have also been very useful in mammals tissues. The previously mentioned gold mine also affected balochistan gerbils, which showed several histological manifestations in liver, kidney, and lung likely related to elevated levels of diverse metals (Cd, Pb, Hg, Zn, Cu, Fe, As, and V) (Almalki et al. 2019b). Other small mammals, such as wild rats and Algerian mice, showed pronounced histopathological damage in kidneys, liver, and spleen, with some of them correlating with high levels of Cd and As (Pereira et al. 2006). Several lesions were also detected in various small mammals species, including edema, intranuclear inclusion bodies, and mitochondrial abnormalities in kidneys. These symptoms of clinical plumbism appeared in animals from areas that showed the highest Pb levels (Roberts et al. 1978). Greater white-toothed shrew also showed a significant increase in number and severity of liver alterations in a mining area in comparison with the reference site. However, no significant relationships were found between pathologies and hepatic metal levels (Sánchez-Chardi et al. 2008). Shrews inhabiting an area affected by the Aznalcóllar spill showed hepatic alterations (higher liver-body ratio, focal necrosis, and signs of apoptosis in hepatocytes) as well as high levels of Pb, Cd, and Hg. However, alterations only correlated with hepatic levels of Pb, Cr, and Mn (Sánchez-Chardi et al. 2009). In Artic hares (*Lepus arcticus*), animals from both former mine and control areas showed histopathological changes in liver and kidney. However, only hares with relatively higher accumulation of Pb from the mine area showed more severe alterations such as renal edema and hemorrhage of capsular surface (Amuno et al. 2016). Similarly, wild snowshoe hares studied near the abandoned Giant mine site exhibited hepatic steatosis (fatty liver change) in both polluted and non-polluted areas, but no other lesions were found (Amuno et al. 2018b). White-footed mice from Southeast Missouri Lead Mining District did not show any lesions associated with Pb or Cd toxicity, despite showing evidence of oxidative stress and reduced ALAD (Beyer et al. 2018). Persian jirds (*Meriones persicus*) chronically exposed to high concentrations of several metals showed numerous histopathological changes in renal and hepatic tissue, but no reference area for comparison was included in the study (Shahsavari et al. 2019). Besides small mammals, goats were used as sentinels of potential histopathological changes caused by indiscriminate small-scale artisanal Au mining activities in Nigeria. Goats were examined for histopathology lesions in the brain. Pb and Cd levels were significantly higher in field exposed goats than in the control group, and those exposed goats showed chromatolysis and increased astrocytic activity in brain tissues (Jubril et al. 2019).

As stated at the beginning of this section, histopathological analyses are very useful to assess metal-related lesions in different organs. However, as these biomarkers required the death of the individual, they should only be used in non-endangered species or in necropsies of dead individuals. Also, even in non-protected species, we must select species with stable populations in which the death of some individuals would not affect the population status.

Other biomarkers related to general health status. Some of the reviewed studies examined parameters that were difficult to be included in other biomarker categories. Because of that, we summarize some of them here. Despite being less widely used, some of them proved to be useful as biomarkers of effect.

Bone pathologies were assessed in 4 studies. White stork nestlings born in the surrounding of the Aznalcóllar mining spillage showed gross pathology of the legs, which presented deformities. However, these deformities were only partially explained by metals bone concentrations (Smits et al. 2005). In addition, these nestlings with leg deformities were unable to control serum phosphorous (P) levels and Ca:P ratios (Smits et al. 2007). Also, changes in bone tissue composition and mineralization were observed in red deer and wild boar associated with Pb levels (Rodríguez-Estival et al. 2013a). Finally, snowshoe hares chronically exposed to As and Cd showed various bone abnormalities in femur and vertebrae, including osteoporosis (Amuno et al. 2018a). A different parameter measured in two small mammals was microbial communities in the small intestines. Microbial assemblages were different in response to metal contamination, and that could have contributed to the overall decline in animal health (Coolon et al. 2010).

Endocrine parameters have proved useful in the detection of early of low-level responses to pollutants, although little is known about the effects of sublethal metal exposure on the adrenal and thyroid systems in free-living birds. Thus, adrenocortical response to stress and thyroid hormone status were assessed in white storks inhabiting an area affected by the Aznalcóllar mine spillage. Maximum levels of corticosterone (stress-induced) were positively correlated to Pb in both control and polluted area. However, thyroid hormones were not related to metal concentrations, although thyroid status differed with location (Baos et al. 2006a).

Neuromotor deficits are an important sign of Mn toxicity in humans and laboratory animals, but impacts of Mn exposure on wild animals remain unknown. The impact of chronic exposure to Mn from an active mine on the motor function was assessed in the marsupial northern quoll (*Dasyurus hallucatus*) by different tests (Amir Abdul Nasir et al. 2018a). Despite they found elevated Mn body burden, performance in motor tests did not diminish. However, quolls with higher Mn levels showed slower speeds approaching a turn, which may reduce success at catching prey and avoiding predators.

Population and Community Level Biomarkers

Some of the reviewed studies (22) also assessed the effects of mining contamination at population level and community level, including reproductive performance through different measurements, species and community diversity, sex ratio, or even genetic diversity.

Population level biomarkers. Several parameters could be affected by mining pollution, having a long-term effect on population dynamics. 13 studies evaluated parameters related to reproduction performance in different bird species and only in one small mammal species. For example, reproduction of white stork population inhabiting near the Aznalcóllar accident was affected by the spillage. A seven-year study showed that female storks grown up in exposed areas experienced a premature breeding senescence compared with females non-developmentally exposed. In addition, this premature senescence was preceded by an unusually high productivity in their early reproductive life. Consequently, these long-term reproductive effects observed could potentially have an impact on population dynamics (Baos et al. 2012). Potential effects of the Aznalcóllar spillage were also assessed in the booted eagle (*Hieraaetus pennatus*) population of Doñana National Park, using long-term data (1976–2000). Mean population fecundity (i.e. fledglings/pairs in the population) decreased after the accident. However, decrease in fecundity over time was also related to the increase of breeding pairs, suggesting density-dependent regulation of the population, and thus nullifying mining spillage effect (Gil-Jiménez et al. 2017). Size of clutches and nestling mortality was used for evaluating the effects of Pb pollution on pied flycatcher breeding. Flycatchers showed small clutches and higher mortality of nestlings, as well as poorer health status in those who survived in mining zones (Berglund et al. 2010). Pb levels also affected tree swallow's probability of daily eggs survival in a mining-impacted stream. It was significantly reduced at the sites with highest Pb levels, although the authors did not find any correlation between average Pb in liver and the probability of daily eggs survival on a colony basis (Custer et al. 2002a). Another parameter, the proportion of eggs hatched, was studied in tree swallows and house wrens (*Troglodytes aedon*), which showed elevated Hg levels in eggs and liver. The percentage of eggs hatched was 70–74%, below the average in the USA ($\geq 85\%$) (Custer et al. 2007). Productivity was also affected in Western grebes (*Aechmophorus occidentalis*) and Clark's grebes (*Aechmophorus clarkia*). Both species had lower productivity in all years of the study in the contaminated lake, but it is unlikely that only the Hg levels found were causing the reduction (Elbert and Anderson 1998). A study assessing reproductive effects on snowy egrets and black-crowned night-herons inhabiting a Hg contaminated area in Lower Carson River basin showed no conclusive evidence of Hg-related depressed hatchability (Henny et al. 2002). After that, they performed a long-term study (10 years) that showed different responses of snowy egrets and black-crowned nigh-herons to Hg pollution in wet and severe drought years. They evaluated reproductive effects considering 0.80 µg/g ww of MeHg in eggs as a reference value that could affect reproduction. Thus, during drought years snowy egret nests with eggs above that value all failed. However, in wet years, both nests above and below that threshold value produced substantial number of young egrets, although nests above that level produced fewer chicks. Regarding nigh-herons, they did not find any evidence of reproductive toxicity to them, so it proves to be a less sensitive species than snowy egret (Hill et al. 2008). A subsequent study evaluated if these young Hg-exposed egrets would be further compromised after leaving the nest. Using telemetry data, dead bird counts and survival data until 90–110 days of age, they found no evidence that elevated Hg compromised survival in the contaminated site versus the reference site (Henny et al. 2017). Effects on reproductive performance can be different for two distinct species inhabiting the same reclaimed mine, such as the mountain bluebird (*Sialia currucoides*) and the tree swallow. The first one initiated clutches later, and one year showed lower fledging success. The second, although also bred later, had a performance equal or better than in other sites (O'Brien and Dawson 2016). Finally, some bird species did not show any effects, as American dipper. Despite having higher levels of methylmercury (MeHg) in eggs in some areas, all populations along all streams had an excellent reproductive success in comparison with other studies (Henny et al. 2005). Also, the bald eagle (*Haliaeetus leucocephalus*) population nesting in Pinchi Lake was not apparently affected by elevated Hg exposure, as reproductive success and productivity were not different from those on reference lakes (Weech et al. 2006). The only study of reproductive performance in small mammals assessed white-footed mice inhabiting a site contaminated by metals. However, no direct link was found between the contamination and alterations of population parameters, such as population size, reproductive success, and others (Phelps and Mcbee 2010).

Another possible effect at population level is a variation in sex ratio, as was observed in a population of white-tailed ptarmigan inhabiting a mine site. The population exhibited a skewed female:male sex ratio of 3:7, due to the higher effects of Cd contamination on female individuals (Larison et al. 2000). Sex ratio was also affected in reptiles, such as the giant sungazer lizard. Individuals from the most contaminated site, which showed the greatest levels of various elements, had an adult sex ratio deviating from 1:1, in favor of female lizard (McIntyre and Whiting 2012). Red-eared slider turtles living in a mining site were also affected. The mining population showed a female-biased sex ratio, while natural populations are expected to show male-biased ratios as adults (Hays and McBee 2010).

Genetic and genomic data were also used as effect biomarkers in three studies. From the point of view of population genetic diversity, deleterious effects were evaluated in plateau mice chronically exposed to mine tailings. The most polluted population was the most genetically distant from the others, and there was a correlation between metal bioaccumulation and reduced population genetic diversity (Mussali-Galante et al. 2013a). Populations of Algerian mice inhabiting two mining sites and a control site showed little genetic differentiation in two mitochondrial markers, the gene cytochrome b (Cytb) and the control region (CR). However, mining populations were each one differentiated from the other two populations in Cytb in one mine and in CR in the other. Also, diversity of each marker within each site differed among mining sites and reference site. All these findings may be explained by demographic changes and increased mutation rates (Quina et al. 2019). Finally, genomic SNP data was used to examine the potential for an evolutionary response to long-term Pb exposure in House sparrow (*Passer domesticus*). They examined 11 localities including two mining sites with elevated levels of Pb contamination. Results showed some differences in loci that could be associated with genes relevant to Pb exposure, such as two metal (Pb and Zn) transporters across cell membranes (Andrew et al. 2019).

Community level biomarkers. Species and community diversities were also assessed in small mammals. A rodent community showed higher diversity at reference than contaminated site. This decrease was probably a response from communities towards metals (Espinosa-Reyes et al. 2014). Other study assessing small mammal species showed lower species diversity at the contaminated site, including lower richness and evenness. In addition this contaminated population was dominated by a single species, the white-footed deer mice (Phelps and McBee 2009). Finally, a study developed in the fragmented Upper Guinean rainforest (Ghana) considered not the contamination by metals, but the impact on rainforest birds of gold mining matrix in a fragmented forest. They showed that the increase of mining surface produced a decrease in populations of forest-dependent birds as well as a decrease in the most specialists species diversity, therefore causing an impact at population and community levels (Deikumah et al. 2014).

Studies including biomarkers at higher levels of biological organization, as population and community levels, showed that those metal-related effects produced at molecular, cellular, or individual levels are certainly affecting wildlife populations. This kind of studies are highly valuable and necessary to assess the future of wildlife species inhabiting or breeding in or near metalliferous mining sites, but also to evaluate the risk of human populations living in those areas. One of the difficulties of assessing effects at population level is that long-term studies are usually required, with the resulting economic cost. In addition, at these levels of biological organization it is more difficult to establish direct links with metal concentrations.

4 Conclusions

Despite being recognized as one of the main sources of global environmental pollution, the existing scientific information about metal ore mining and its effects on terrestrial and semi-terrestrial wildlife is scarce. The number of studies in active mine sites was much lower than in old mining sites. We thought this could be explained by the current legislation regarding metalliferous mine activities and by the great problem that derelict mines represent for human and environmental health even nowadays. Their danger is associated both with the high number of these orphan mines around the world which have been ignored until recently and with the great amount of bioavailable metals that they exhibit. Also, the low number of studies in countries which are currently among the top metal producers, such as China, Chile, or Australia, should be alarming. And especially, in those developing countries such as Zambia or Ghana, in which legislation is still scarce. It is also noteworthy that all studies assessing effects of a mining spill were carried out at the same location. Despite the great impact that such events could have on ecosystems, no other scientific publications were found assessing impacts on terrestrial vertebrates. Although reclamation and rehabilitation of mines was not the scope of the review, we observed that only 10 of the 128 studies developed in former mines were in reclaimed, rehabilitated, or revegetated mines. The effectiveness of remediation treatments for reducing metal effects in vertebrate species was only measured in some of them. Usually, after-remediation studies focus on vegetation or in animal species richness, but ignore if there is a real reduction in metals bioavailability and accumulation in resident wildlife. Information about metals exposure and effects in sentinel species must be included in reclamation and rehabilitation protocols.

Regarding sentinel species, great differences were found among the use of different taxa groups. Mammals and birds were by far the most assessed species, and there were two clearly underrepresented groups: reptiles and amphibians. Although this is the general trend found also in other pollutants assessments, this does not make it less worrying, as amphibian and reptilian populations keep decreasing. Some particular species were repeatedly used in several studies, so they proved to be good sentinel species that serve as early warning of risks to wildlife and human's health. Finally, a wide range of different biomarkers were used in the reviewed studies. Some only included biomarkers of exposure but the majority evaluated various biomarkers of effect at different levels of biological organization.

We provide with this review a useful database that could be used by the scientific community and governments to further develop ecotoxicological risk assessments and regulations to protect both wildlife and human health from metalliferous mining-related pollution. Next steps in ecotoxicological assessment may be directed to transform the "single biomarker approach" and establish adverse outcome pathways for the different metals (Ankley et al. 2010). The purpose of this holistic approach is to be able to create links between knowledge at the different levels of biological organization, from the molecular levels to ecosystem levels (Mateo et al. 2016). This requires knowledge of the mechanisms of action of the different elements at molecular level, as well as responses at cellular and organ levels and the effects at individual or population levels. To be able to develop this knowledge, there should be a combination of experimental work on surrogate sentinel species to better understand metal toxicological thresholds (and their confounding factors) with correlational studies on wildlife health using sensitive and reliable biomarkers of exposure and effect (Rodríguez-Estival and Mateo 2019). Ideally, non-destructive and non-invasive techniques should be used to diminish the impact on the studied organisms. And finally, especially for population and ecosystem levels, long-term environmental studies are essential. All progress made in elucidating mechanisms and effects of metals in wildlife would facilitate the inclusion of accurate parameters in protocols of mining-related assessments and legislation, in order to effectively protect both wildlife and human health.

Acknowledgment This work was supported by funding from Fundación Migres. We would like to thank C. Ayora and anonymous referees for their comments that greatly improved a previous draft of this review.

Conflict of Interest Statement On behalf of all authors, the corresponding author states that there is no conflict of interest.

References

Abeysinghe KS, Qiu G, Goodale E et al (2017) Mercury flow through an Asian rice-based food web. Environ Pollut 229:219–228. https://doi.org/10.1016/j.envpol.2017.05.067

Almalki AM, Ajarem J, Allam AA et al (2019a) Use of spilopelia senegalensis as a biomonitor of heavy metal contamination from mining activities in riyadh (Saudi Arabia). Animals 9:1–18. https://doi.org/10.3390/ani9121046

Almalki AM, Ajarem J, Altoom N et al (2019b) Effects of mining activities on Gerbillus nanus in Saudi Arabia: a biochemical and histological study. Animals 9:1–15. https://doi.org/10.3390/ani9090664

Almli B, Mwase M, Sivertsen T et al (2005) Hepatic and renal concentrations of 10 trace elements in crocodiles (Crocodylus niloticus) in the Kafue and Luangwa rivers in Zambia. Sci Total Environ 337:75–82. https://doi.org/10.1016/j.scitotenv.2004.06.019

Alvarenga P, Simões I, Palma P et al (2014) Field study on the accumulation of trace elements by vegetables produced in the vicinity of abandoned pyrite mines. Sci Total Environ 470–471:1233–1242. https://doi.org/10.1016/j.scitotenv.2013.10.087

Amir Abdul Nasir AF, Cameron SF, Niehaus AC et al (2018a) Manganese contamination affects the motor performance of wild northern quolls (Dasyurus hallucatus). Environ Pollut 241:55–62. https://doi.org/10.1016/j.envpol.2018.03.087

Amir Abdul Nasir AF, Cameron SF, von Hippel FA et al (2018b) Manganese accumulates in the brain of northern quolls (Dasyurus hallucatus) living near an active mine. Environ Pollut 233:377–386. https://doi.org/10.1016/j.envpol.2017.10.088

Amuno S, Niyogi S, Amuno M, Attitaq J (2016) Heavy metal bioaccumulation and histopathological alterations in wild Arctic hares (Lepus arcticus) inhabiting a former lead-zinc mine in the Canadian high Arctic: a preliminary study. Sci Total Environ 556:252–263. https://doi.org/10.1016/j.scitotenv.2016.03.007

Amuno S, Al Kaissi A, Jamwal A et al (2018a) Chronic arsenicosis and cadmium exposure in wild snowshoe hares (Lepus americanus) breeding near Yellowknife, Northwest Territories (Canada), part 2: manifestation of bone abnormalities and osteoporosis. Sci Total Environ 612:1559–1567. https://doi.org/10.1016/j.scitotenv.2017.08.280

Amuno S, Jamwal A, Grahn B, Niyogi S (2018b) Chronic arsenicosis and cadmium exposure in wild snowshoe hares (Lepus americanus) breeding near Yellowknife, Northwest Territories (Canada), part 1: evaluation of oxidative stress, antioxidant activities and hepatic damage. Sci Total Environ 618:916–926. https://doi.org/10.1016/j.scitotenv.2017.08.278

Andrew SC, Taylor MP, Lundregan S et al (2019) Signs of adaptation to trace metal contamination in a common urban bird. Sci Total Environ 650:679–686. https://doi.org/10.1016/j.scitotenv.2018.09.052

Andrews SM, Johnson MS, Cooke JA (1984) Cadmium in small mammals from grassland established on metalliferous mine waste. Environ Pollut (Series A) 33:153–162

Ankley GT, Bennett RS, Erickson RJ et al (2010) Adverse outcome pathways: a conceptual framework to support ecotoxicology research and risk assessment. Environ Toxicol Chem 29:730–741. https://doi.org/10.1002/etc.34

ATSDR (2019) Agency for toxic substances and disease registry. Substance Priority List. https://www.atsdr.cdc.gov/SPL/index.html

Avery RA, White AS, Martin MH, Hopkin SP (1983) Concentrations of heavy metals in common lizards (Lacerta vivipara) and their food and environment. Amphibia-Reptilia 4:205–213

Baos R, Blas J, Bortolotti GR et al (2006a) Adrenocortical response to stress and thyroid hormone status in free-living nestling white storks (Ciconia ciconia) exposed to heavy metal and arsenic contamination. Environ Health Perspect 114:1497–1501. https://doi.org/10.1289/ehp.9099

Baos R, Jovani R, Forero MG et al (2006b) Relationships between T-cell-mediated immune response and Pb, Zn, Cu, Cd, and As concentrations in blood of nestling white storks (Ciconia ciconia) and black kites (Milvus migrans) from Doñana (southwestern Spain) after the

Aznalcóllar toxic spill. Environ Toxicol Chem 25:1153–1159. https://doi.org/10.1897/05-395R.1

Baos R, Jovani R, Pastor N et al (2006c) Evaluation of genotoxic effects of heavy metals and arsenic in wild nestling white storks (Ciconia ciconia) and black kites (Milvus migrans) from southwestern Spain after a mining accident. Environ Toxicol Chem 25:2794–2803. https://doi.org/10.1897/05-570r.1

Baos R, Jovani R, Serrano D et al (2012) Developmental exposure to a toxic spill compromises long-term reproductive performance in a wild, long-lived bird: the white stork (Ciconia ciconia). PLoS One 7:e34716. https://doi.org/10.1371/journal.pone.0034716

Barata C, Fabregat MC, Cotín J et al (2010) Blood biomarkers and contaminant levels in feathers and eggs to assess environmental hazards in heron nestlings from impacted sites in Ebro basin (NE Spain). Environ Pollut 158:704–710. https://doi.org/10.1016/j.envpol.2009.10.018

Basu N, Scheuhammer AM, Bursian SJ et al (2007) Mink as a sentinel species in environmental health. Environ Res 103:130–144. https://doi.org/10.1016/j.envres.2006.04.005

Benito V, Devesa V, Muñoz O et al (1999) Trace elements in blood collected from birds feeding in the area around Doñana National Park affected by the toxic spill from the Aznalcóllar mine. Sci Total Environ 242:309–323. https://doi.org/10.1016/S0048-9697(99)00398-8

Benson WW, Brock DW, Gabica J, Loomis M (1976) Swan mortality due to certain heavy metals in the Mission Lake Area, Idaho. Bull Environ Contam Toxicol 15:171–174. https://doi.org/10.1007/BF01685156

Berglund ÅMM (2018) Evaluating blood and excrement as bioindicators for metal accumulation in birds. Environ Pollut 233:1198–1206. https://doi.org/10.1016/j.envpol.2017.10.031

Berglund AMM, Sturve J, Förlin L, Nyholm NEI (2007) Oxidative stress in pied flycatcher (Ficedula hypoleuca) nestlings from metal contaminated environments in northern Sweden. Environ Res 105:330–339. https://doi.org/10.1016/j.envres.2007.06.002

Berglund ÅMM, Ingvarsson PK, Danielsson H, Nyholm NEI (2010) Lead exposure and biological effects in pied flycatchers (Ficedula hypoleuca) before and after the closure of a lead mine in northern Sweden. Environ Pollut 158:1368–1375. https://doi.org/10.1016/j.envpol.2010.01.005

Berglund ÅMM, Koivula MJ, Eeva T (2011) Species- and age-related variation in metal exposure and accumulation of two passerine bird species. Environ Pollut 159:2368–2374. https://doi.org/10.1016/j.envpol.2011.07.001

Berzas Nevado JJ, Rodríguez Martin-Doimeadios RC, Mateo R et al (2012) Mercury exposure and mechanism of response in large game using the Almadén mercury mining area (Spain) as a case study. Environ Res 112:58–66. https://doi.org/10.1016/j.envres.2011.09.019

Beyer WN, Spann JW, Sileo L, Franson JC (1988) Lead poisoning in six captive avian species. Arch Environ Contam Toxicol 17:121–130. https://doi.org/10.1007/BF01055162

Beyer WN, Audet DJ, Morton A et al (1998) Lead Exposure of waterfowl ingesting Coeur d'Alene River Basin sediments. J Environ Qual 27:1533–1538

Beyer WN, Dalgarn J, Dudding S et al (2004) Zinc and lead poisoning in wild birds in the Tri-state mining district (Oklahoma, Kansas, and Missouri). Arch Environ Contam Toxicol 48:108–117. https://doi.org/10.1007/s00244-004-0010-7

Beyer WN, Gaston G, Brazzle R et al (2007) Deer exposed to exceptionally high concentrations of lead near the Continental Mine in Idaho, USA. Environ Toxicol Chem 26:1040–1046. https://doi.org/10.1897/06-304R.1

Beyer WN, Franson JC, French JB et al (2013) Toxic exposure of songbirds to lead in the Southeast Missouri Lead Mining District. Arch Environ Contam Toxicol 65:598–610. https://doi.org/10.1007/s00244-013-9923-3

Beyer WN, Casteel SW, Friedrichs KR et al (2018) Biomarker responses of Peromyscus leucopus exposed to lead and cadmium in the Southeast Missouri Lead Mining District. Environ Monit Assess:190. https://doi.org/10.1007/s10661-017-6442-0

Blus LJ, Henny CJ (1990) Lead and cadmium concentrations in mink from northern Idaho. Northwest Sci 64:219–223

Blus LJ, Henny CJ, Anderson A, Fitzner RE (1985) Reproduction, mortality, and heavy metal concentrations in Great Blue Herons from three colonies in Washington and Idaho. Colon Waterbirds 8:110–116

Blus LJ, Henny CJ, Mulhern BM (1987) Concentrations of metals in mink and other mammals from Washington and Idaho. Environ Pollut 44:307–318. https://doi.org/10.1016/0269-7491(87)90206-5

Blus LJ, Henny CJ, Hoffman DJ, Grove RA (1991) Lead toxicosis in tundra swans near a mining and smelting complex in northern Idaho. Arch Environ Contam Toxicol 21:549–555. https://doi.org/10.1007/BF01183877

Blus LJ, Henny CJ, Hoffman DJ, Grove RA (1993) Accumulation and effects of lead and cadmium on wood ducks near a mining and smelting complex in Idaho. Ecotoxicology 2:139–154. https://doi.org/10.1007/BF00119436

Blus LJ, Henny CJ, Hoffman DJ, Grove RA (1995) Accumulation in and effects of lead and cadmium on waterfowl and passerines in northern Idaho. Environ Pollut 89:311–318

Blus LJ, Henny CJ, Hoffman DJ et al (1999) Persistence of high lead concentrations and associated effects in Tundra Swans captured near a mining and smelting complex in Northern Idaho. Ecotoxicology 8:125–132

Bonilla-Valverde D, Ruiz-Laguna J, Muñoz A et al (2004) Evolution of biological effects of Aznalcóllar mining spill in the Algerian mouse (Mus spretus) using biochemical biomarkers. Toxicology 197:123–138. https://doi.org/10.1016/j.tox.2003.12.010

Bortey-Sam N, Nakayama SMM, Ikenaka Y et al (2015) Human health risks from metals and metalloid via consumption of food animals near gold mines in Tarkwa, Ghana: estimation of the daily intakes and target hazard quotients (THQs). Ecotoxicol Environ Saf 111:160–167. https://doi.org/10.1016/j.ecoenv.2014.09.008

Bortey-Sam N, Nakayama SMM, Ikenaka Y et al (2016) Heavy metals and metalloid accumulation in livers and kidneys of wild rats around gold-mining communities in Tarkwa, Ghana. J Environ Chem Ecotoxicol 8:58–68. https://doi.org/10.5897/jece2016.0374

Brumbaugh WG, Mora MA, May TW, Phalen DN (2010) Metal exposure and effects in voles and small birds near a mining haul road in Cape Krusenstern National Monument, Alaska. Environ Monit Assess 170:73–86. https://doi.org/10.1007/s10661-009-1216-y

Bureau of Land Management (2019). Abandonedmines.gov. https://www.abandonedmines.gov/

Burger J (2006) Bioindicators: a review of their use in the environmental literature 1970–2005. Environ Bioindic 1:136–144. https://doi.org/10.1080/15555270600701540

Burger J (2008) Assessment and management of risk to wildlife from cadmium. Sci Total Environ 389:37–45. https://doi.org/10.1016/j.scitotenv.2007.08.037

Camizuli E, Scheifler R, Garnier S et al (2018) Trace metals from historical mining sites and past metallurgical activity remain bioavailable to wildlife today. Sci Rep 8:1–11. https://doi.org/10.1038/s41598-018-20983-0

Carpenter JW, Andrews GA, Beyer WN (2004) Zinc toxicosis in a free-flying trumpeter swan (Cygnus buccinator). J Wildl Dis 40:769–774. https://doi.org/10.7589/0090-3558-40.4.769

Carvalho FP (2017) Mining industry and sustainable development: time for change. Food Energy Secur 6:61–77. https://doi.org/10.1002/fes3.109

Castellanos P, Reglero MM, Taggart MA, Mateo R (2010) Changes in fatty acid profiles in testis and spermatozoa of red deer exposed to metal pollution. Reprod Toxicol 29:346–352. https://doi.org/10.1016/j.reprotox.2010.01.005

Castellanos P, Del Olmo E, Fernández-Santos MR et al (2015) Increased chromatin fragmentation and reduced acrosome integrity in spermatozoa of red deer from lead polluted sites. Sci Total Environ 505:32–38. https://doi.org/10.1016/j.scitotenv.2014.09.087

Chaousis S, Leusch FDL, van de Merwe JP (2018) Charting a path towards non-destructive biomarkers in threatened wildlife: a systematic quantitative literature review. Environ Pollut 234:59–70. https://doi.org/10.1016/j.envpol.2017.11.044

Chapa-Vargas L, Mejia-Saavedra JJ, Monzalvo-Santos K, Puebla-Olivares F (2010) Blood lead concentrations in wild birds from a polluted mining region at Villa de La Paz, San Luis Potosi,

Mexico. J Environ Sci Health A Tox Hazard Subst Environ Eng 45:90–98. https://doi.org/10.1080/10934520903389242

Conder JM, Lanno RP (1999) Heavy metal concentrations in mandibles of white-tailed deer living in the Picher Mining District. Bull Environ Contam Toxicol 63:80–86

Coolon JD, Jones KL, Narayanan S, Wisely SM (2010) Microbial ecological response of the intestinal flora of Peromyscus maniculatus and P. leucopus to heavy metal contamination. Mol Ecol 19:67–80. https://doi.org/10.1111/j.1365-294X.2009.04485.x

Costa RA, Petronilho JMS, Soares AMVM, Vingada JV (2011) The use of passerine feathers to evaluate heavy metal pollution in Central Portugal. Bull Environ Contam Toxicol 86:352–356. https://doi.org/10.1007/s00128-011-0212-4

Cristescu RH, Frère C, Banks PB (2012) A review of fauna in mine rehabilitation in Australia: current state and future directions. Biol Conserv 149:60–72. https://doi.org/10.1016/j.biocon.2012.02.003

Custer CM (2011) Swallows as a sentinel species for contaminant exposure and effect studies. In: Elliott JE, Bishop CA, Morrissey CA (eds) Wildlife ecotoxicology: forensic approaches. Springer, New York, pp 45–91

Custer CM, Custer TW, Archuleta AS et al (2002a) A mining impacted stream: exposure and effects of lead and other trace elements on tree swallows (Tachycineta bicolor) nesting in the Upper Arkansas River Basin, Colorado. In: Hoffman DJ, Rattner BA, Burton GA Jr, Cairns J Jr (eds) Handbook of ecotoxicology, 2nd edn. CRC Press

Custer TW, Custer CM, Larson S, Dickerson KK (2002b) Arsenic concentrations in house wrens from Whitewood Creek, South Dakota, USA. Bull Environ Contam Toxicol 68:517–524. https://doi.org/10.1007/s001280285

Custer CM, Custer TW, Hill EF (2007) Mercury exposure and effects on cavity-nesting birds from the Carson River, Nevada. Arch Environ Contam Toxicol 52:129–136. https://doi.org/10.1007/s00244-006-0103-6

Custer CM, Yang C, Crock JG et al (2009) Exposure of insects and insectivorous birds to metals and other elements from abandoned mine tailings in three Summit County drainages, Colorado. Environ Monit Assess 153:161–177. https://doi.org/10.1007/s10661-008-0346-y

DalCorso G (2012) Chapter 1: heavy metal toxicity in plants. In: Furini A (ed) Plants and heavy metals. Springer, Dordrecht, pp 1–25

Damek-Poprawa M, Sawicka-Kapusta K (2003) Damage to the liver, kidney, and testis with reference to burden of heavy metals in yellow-necked mice from areas around steelworks and zinc smelters in Poland. Toxicology 186:1–10

Damek-Poprawa M, Sawicka-Kapusta K (2004) Histopathological changes in the liver, kidneys, and testes of bank voles environmentally exposed to heavy metal emissions from the steelworks and zinc smelter in Poland. Environ Res 96:72–78. https://doi.org/10.1016/j.envres.2004.02.003

Deikumah JP, McAlpine CA, Maron M (2014) Mining matrix effects on West African rainforest birds. Biol Conserv 169:334–343. https://doi.org/10.1016/j.biocon.2013.11.030

Delibes M, Cabezas S, Jiménez B, González MJ (2009) Animal decisions and conservation: the recolonization of a severely polluted river by the Eurasian otter. Anim Conserv 12:400–407. https://doi.org/10.1111/j.1469-1795.2009.00263.x

Di Marzio A, Lambertucci SA, Fernandez AJG, Martínez-López E (2019) From Mexico to the Beagle Channel: a review of metal and metalloid pollution studies on wildlife species in Latin America. Environ Res 176:108462. https://doi.org/10.1016/j.envres.2019.04.029

Doganoc DZ, Gačnik KS (1995) Lead and cadmium in meat and organs of game in Slovenia. Bull Environ Contam Toxicol 54:166–170. https://doi.org/10.1007/BF00196284

Doumas P, Munoz M, Banni M et al (2018) Polymetallic pollution from abandoned mines in Mediterranean regions: a multidisciplinary approach to environmental risks. Reg Environ Chang 18:677–692. https://doi.org/10.1007/s10113-016-0939-x

Drouhot S, Raoul F, Crini N et al (2014) Responses of wild small mammals to arsenic pollution at a partially remediated mining site in Southern France. Sci Total Environ 470–471:1012–1022. https://doi.org/10.1016/j.scitotenv.2013.10.053

Elbert RA, Anderson DW (1998) Mercury levels, reproduction, and hematology in Western grebes from three California lakes, USA. Environ Toxicol Chem 17:210–213. https://doi.org/10.1002/etc.5620170212

Erry BV, Macnair MR, Meharg AA et al (1999a) Arsenic residues in predatory birds from an area of Britain with naturally and anthropogenically elevated arsenic levels. Environ Pollut 106:91–95

Erry BV, Macnair MR, Meharg AA, Shore RF (1999b) Seasonal variation in dietary and body organ arsenic concentrations in wood mice Apodemus sylvaticus and bank voles Clethrionomys glareolus. Bull Environ Contam Toxicol 63:567–574

Erry BV, Macnair MR, Meharg AA, Shore RF (2000) Arsenic contamination in wood mice (Apodemus sylvaticus) and bank voles (Clethrionomys glareolus) on abandoned mine sites in southwest Britain. Environ Pollut 110:179–187

Erry BV, Macnair MR, Meharg AA, Shore RF (2005) The distribution of arsenic in the body tissues of wood mice and bank voles. Arch Environ Contam Toxicol 49:569–576. https://doi.org/10.1007/s00244-004-0229-3

Espín S, Martínez-López E, León-Ortega M et al (2014a) Oxidative stress biomarkers in Eurasian eagle owls (Bubo bubo) in three different scenarios of heavy metal exposure. Environ Res 131:134–144. https://doi.org/10.1016/j.envres.2014.03.015

Espín S, Martínez-López E, León-Ortega M et al (2014b) Factors that influence mercury concentrations in nestling Eagle Owls (Bubo bubo). Sci Total Environ 470–471:1132–1139. https://doi.org/10.1016/j.scitotenv.2013.10.063

Espín S, Martínez-López E, Jiménez P et al (2015) Delta-aminolevulinic acid dehydratase (δALAD) activity in four free-living bird species exposed to different levels of lead under natural conditions. Environ Res 137C:185–198. https://doi.org/10.1016/j.envres.2014.12.017

Espinosa-Reyes G, Torres-Dosal A, Ilizaliturri C et al (2010) Wild rodents (Dipodomys merriami) used as biomonitors in contaminated mining sites. J Environ Sci Health A Tox Hazard Subst Environ Eng 45:82–89. https://doi.org/10.1080/10934520903388988

Espinosa-Reyes G, González-Mille DJ, Ilizaliturri-Hernández CA et al (2014) Effect of mining activities in biotic communities of Villa de la Paz, San Luis Potosi, Mexico. Biomed Res Int 2014:165046. https://doi.org/10.1155/2014/165046

European Commission (2017) Closed and abandoned waste facilities. National Inventories. https://ec.europa.eu/environment/waste/mining/implementation.htm

Festa F, Cristaldi M, Ieradi LA et al (2003) The Comet assay for the detection of DNA damage in Mus spretus from Doñana National Park. Environ Res 91:54–61. https://doi.org/10.1016/s0013-9351(02)00003-8

Fletcher DE, Hopkins WA, Saldaña T et al (2006) Geckos as indicators of mining pollution. Environ Toxicol Chem 25:2432–2445. https://doi.org/10.1897/05-556R.1

Franson JC, Pain DJ (2011) Lead in birds. In: Beyer WN, Meador JP (eds) Enviromental contaminants in biota. Interpreting tissue concentraions, 2nd edn. CRC Press, Taylor & Francis Group, Boca Raton, pp 563–594

Fu Z, Wu F, Mo C et al (2011) Bioaccumulation of antimony, arsenic, and mercury in the vicinities of a large antimony mine, China. Microchem J 97:12–19. https://doi.org/10.1016/j.microc.2010.06.004

Garitano-Zavala Á, Cotín J, Borràs M, Nadal J (2010) Trace metal concentrations in tissues of two tinamou species in mining areas of Bolivia and their potential as environmental sentinels. Environ Monit Assess 168:629–644. https://doi.org/10.1007/s10661-009-1139-7

Gil F, Capitán-Vallvey LF, De Santiago E et al (2006) Heavy metal concentrations in the general population of Andalusia, South of Spain: a comparison with the population within the area of influence of Aznalcóllar mine spill (SW Spain). Sci Total Environ 372:49–57. https://doi.org/10.1016/j.scitotenv.2006.08.004

Gil-Jiménez E, Manzano J, Casado E, Ferrer M (2017) The role of density-dependence regulation in the misleading effect of the Aznalcollar mining spill on the booted eagle fecundity. Sci Total Environ 583:440–446. https://doi.org/10.1016/j.scitotenv.2017.01.098

Gómez G, Baos R, Gómara B et al (2004) Influence of a mine tailing accident near Doñana National Park (Spain) on heavy metals and arsenic accumulation in 14 species of waterfowl (1998 to 2000). Arch Environ Contam Toxicol 47:521–529. https://doi.org/10.1007/s00244-004-0189-z

Gómez-Ramírez P, Martínez-López E, María-Mojica P et al (2011) Blood lead levels and δ-ALAD inhibition in nestlings of Eurasian Eagle Owl (Bubo bubo) to assess lead exposure associated to an abandoned mining area. Ecotoxicology 20:131–138. https://doi.org/10.1007/s10646-010-0563-3

Gómez-Ramírez P, Shore RF, van den Brink NW et al (2014) An overview of existing raptor contaminant monitoring activities in Europe. Environ Int 67:12–21. https://doi.org/10.1016/j.envint.2014.02.004

Green Cross Switzerland, Pure Earth (2016) World's worst pollution problems – the toxics beneath our feet

Grillitsch B, Schiesari L (2010) The ecotoxicology of metals in reptiles. In: Sparling DW, Linder G, Bishop CA, Krest SK (eds) Ecotoxicology of amphibians and reptiles, second. CRC Press, Taylor & Francis Group, Boca Raton, pp 337–448

Grimalt JO, Ferrer M, Macpherson E (1999) The mine tailing accident in Aznalcollar. Sci Total Environ 242:3–11. https://doi.org/10.1016/S0048-9697(99)00372-1

Gutiérrez M, Mickus K, Camacho LM (2016) Abandoned Pb-Zn mining wastes and their mobility as proxy to toxicity: a review. Sci Total Environ 565:392–400. https://doi.org/10.1016/j.scitotenv.2016.04.143

Hansen JJ, Audet D, Spears BB et al (2011) Lead exposure and poisoning of songbirds using the Coeur d'Alene River Basin, Idaho, USA. Integr Environ Assess Manag 7:587–595. https://doi.org/10.1002/ieam.201

Hays KA, McBee K (2007) Flow cytometric analysis of red-eared slider turtles (Trachemys scripta) from Tar Creek Superfund site. Ecotoxicology 16:353–361. https://doi.org/10.1007/s10646-007-0135-3

Hays KA, McBee K (2010) Population demographics of red-eared slider turtles (Trachemys scripta) from Tar Creek Superfund site. J Herpetol 44:441–446

He C, Su T, Liu S et al (2020) Heavy metal, arsenic, and selenium concentrations in bird feathers from a region in Southern China impacted by intensive mining of nonferrous metals. Environ Toxicol Chem 39:371–380. https://doi.org/10.1002/etc.4622

Henny CJ, Blus LJ, Hoffman DJ et al (1991) Lead accumulation and osprey production near a mining site on the Coeur d'Alene River, Idaho. Arch Environ Contam Toxicol 21:415–424. https://doi.org/10.1007/BF01060365

Henny CJ, Blus LJ, Hoffman DJ, Grove RA (1994) Lead in hawks, falcons and owls downstream from a mining site on the Coeur d'Alene River, Idaho. Environ Monit Assess 29:267–288. https://doi.org/10.1007/BF00547991

Henny CJ, Blus LJ, Hoffman DJ et al (2000) Field evaluation of lead effects on Canada geese and mallards in the Coeur d'Alene River Basin, Idaho. Arch Environ Contam Toxicol 39:97–112. https://doi.org/10.1007/s002440010085

Henny CJ, Hill EF, Hoffman DJ et al (2002) Nineteenth century mercury: hazard to wading birds and cormorants of the Carson River, Nevada. Ecotoxicology 11:213–231

Henny CJ, Kaiser JL, Packard HA et al (2005) Assessing mercury exposure and effects to American dippers in headwater streams near mining sites. Ecotoxicology 14:709–725. https://doi.org/10.1007/s10646-005-0023-7

Henny CJ, Hill EF, Grove RA, Kaiser JL (2007) Mercury and drought along the lower Carson River, Nevada: I. Snowy egret and black-crowned night-heron annual exposure to mercury, 1997-2006. Arch Environ Contam Toxicol 53:269–280. https://doi.org/10.1007/s00244-006-0163-7

Henny CJ, Hill EF, Grove RA et al (2017) Mercury and drought along the lower Carson River, Nevada: IV. Snowy egret post-fledging dispersal, timing of migration and survival, 2002–2004. Ecotoxicol Environ Saf 135:358–367. https://doi.org/10.1016/j.ecoenv.2016.10.002

Hermoso de Mendoza García M, Soler Rodríguez F, Pérez López M (2008) Los mamíferos salvajes terrestres como bioindicadores: nuevos avances en Ecotoxicología. Obs Medioambient 11:37–62

Hernández L, Gomara B, Fernandez M et al (1999) Accumulation of heavy metals and As in wetland birds in the area around Donana National Park affected by the Aznalcollar toxic spill. Sci Total Environ:293–308

Hill EF, Henny CJ, Grove RA (2008) Mercury and drought along the lower Carson River, Nevada: II. Snowy egret and black-crowned night-heron reproduction on Lahontan Reservoir, 1997-2006. Ecotoxicology 17:117–131. https://doi.org/10.1007/s10646-007-0180-y

Hoffman DJ, Henny CJ, Hill EF et al (2009) Mercury and drought along the lower Carson river, Nevada: III. Effects on blood and organ biochemistry and histopathology of snowy egrets and Black-crowned night-herons on Lahontan reservoir, 2002-2006. J Toxicol Environ Health A 72:1223–1241. https://doi.org/10.1080/15287390903129218

Hopkins WA (2000) Letter to the editor, reptile toxicology: challenges and opportunities on the last front frontier in vertebrate ecotoxicology. Environ Toxicol Chem 19:2391–2393

Hopkins WA, Roe JH, Snodgrass JW et al (2001) Nondestructive indices of trace element exposure in squamate reptiles. Environ Pollut 115:1–7

Ikenaka Y, Nakayama SMM, Muroya T et al (2012) Effects of environmental lead contamination on cattle in a lead/zinc mining area: changes in cattle immune systems on exposure to lead in vivo and in vitro. Environ Toxicol Chem 31:2300–2305. https://doi.org/10.1002/etc.1951

IRP (2020) Mineral resource governance in the 21st century: gearing extractive industries towards sustainable development. Ayuk ET, Pedro AM, Ekins P, Gatune J, Milligan B, Oberle B, Christmann P, Ali S, Kumar SV, Bringezu S, Acquatella J, Bernaudat L, Bodouroglou C, Brooks S, Buergi Bonanomi E, Clement J, Collins N, Davis K, Davy A, Dawkins K, Dom A, Eslamishoar F, Franks D, Hamor T, Jensen D, Lahiri-Dutt K, Mancini L, Nuss P, Petersen I, Sanders ARD. A report by the International Resource Panel. United Nations Environment Programme, Nairobi, Kenya

Isaksson C (2010) Pollution and its impact on wild animals: a meta-analysis on oxidative stress. Ecohealth 7:342–350. https://doi.org/10.1007/s10393-010-0345-7

Jasso-Pineda Y, Espinosa-Reyes G, González-Mille D et al (2007) An integrated health risk assessment approach to the study of mining sites contaminated with arsenic and lead. Integr Environ Assess Manag 3:344–350

Ji X, Hu W, Cheng J et al (2006) Oxidative stress on domestic ducks (Shaoxing duck) chronically exposed in a Mercury-Selenium coexisting mining area in China. Ecotoxicol Environ Saf 64:171–177. https://doi.org/10.1016/j.ecoenv.2005.03.009

Johnson MS, Roberts RD, Hutton M, Inskip MJ (1978) Distribution of lead, zinc and cadmium in small mammals from polluted environments. Oikos 30:153–159. https://doi.org/10.2307/3543536

Johnson GD, Audet DJ, Kern JW et al (1999) Lead exposure in passerines inhabiting lead-contaminated floodplains in the Coeur D'Alene River basin, Idaho, USA. Environ Toxicol Chem 18:1190–1194. https://doi.org/10.1002/etc.5620180617

Jubril AJ, Kabiru M, Olapade JO, Taiwo VO (2017) Biological monitoring of heavy metals in goats exposed to environmental contamination in Bagega. Adv Environ Biol 11:11–18

Jubril AJ, Obasa AA, Mohammed SA et al (2019) Neuropathological lesions in the brains of goats in North-Western Nigeria: possible impact of artisanal mining. Environ Sci Pollut Res 26:36589–36597. https://doi.org/10.1007/s11356-019-06611-y

Junge RE, Williams CV, Rakotondrainibe H et al (2017) Baseline health and nutrition evaluation of two sympatric nocturnal lemur species (Avahi Laniger and Lepilemur Mustelinus) residing near an active mine site at Ambatovy, Madagascar. J Zoo Wildl Med 48:794–803. https://doi.org/10.1638/2016-0261.1

Kakkar P, Jaffery FN (2005) Biological markers for metal toxicity. Environ Toxicol Pharmacol 19:335–349. https://doi.org/10.1016/j.etap.2004.09.003

Kalisińska E, Salicki W, Mysłek P et al (2004) Using the Mallard to biomonitor heavy metal contamination of wetlands in north-western Poland. Sci Total Environ 320:145–161. https://doi.org/10.1016/j.scitotenv.2003.08.014

Khazaee M, Hamidian AH, Alizadeh Shabani A et al (2016) Accumulation of heavy metals and As in liver, hair, femur, and lung of Persian jird (Meriones persicus) in Darreh Zereshk copper mine, Iran. Environ Sci Pollut Res 23:3860–3870. https://doi.org/10.1007/s11356-015-5455-x

Kirsch-Volders M, Plas G, Elhajouji A et al (2011) The in vitro MN assay in 2011: origin and fate, biological significance, protocols, high throughput methodologies and toxicological relevance. Arch Toxicol 85:873–899. https://doi.org/10.1007/s00204-011-0691-4

Klobučar GÖIVV, Pavlica M, Erben R, Papeš D (2003) Application of the micronucleus and comet assays to mussel Dreissena polymorpha haemocytes for genotoxicity monitoring of freshwater environments. Aquat Toxicol 64:15–23. https://doi.org/10.1016/S0166-445X(03)00009-2

Koch I, Mace JV, Reimer KJ (2005) Arsenic speciation in terrestrial birds from Yellowknife, Northwest Territories, Canada: the unexpected finding of arsenobetaine. Environ Toxicol Chem 24:1468–1474. https://doi.org/10.1897/04-155R.1

Koenig R (2000) Wildlife deaths are a grim wake-up call in Eastern Europe. Science 287:1737–1738

Langner HW, Greene E, Domenech R, Staats MF (2012) Mercury and other mining-related contaminants in ospreys along the Upper Clark Fork River, Montana, USA. Arch Environ Contam Toxicol 62:681–695. https://doi.org/10.1007/s00244-011-9732-5

Larison JR, Likens GE, Fitzpatrick JW, Crock JG (2000) Cadmium toxicity among wildlife in the Colorado Rocky Mountains. Nature 406:181–183. https://doi.org/10.1038/35018068

Laurinolli M, Bendell-Young LI (1996) Copper, zinc, and cadmium concentrations in Peromyscus maniculatus sampled near an abandoned copper mine. Arch Environ Contam Toxicol 30:481–486

Lee YH, Stuebing RB (1990) Heavy metal contamination in the River Toad, Bufo juxtasper (Inger), near a copper mine in East Malaysia. Bull Environ Contam Toxicol 45:272–279. https://doi.org/10.1007/BF01700195

Lester MB, van Riper C (2014) The distribution and extent of heavy metal accumulation in song sparrows along Arizona's upper Santa Cruz River. Environ Monit Assess 186:4779–4791. https://doi.org/10.1007/s10661-014-3737-2

Lidman J, Jonsson M, Berglund ÅMM (2020) Availability of specific prey types impact pied flycatcher (Ficedula hypoleuca) nestling health in a moderately lead contaminated environment in northern Sweden. Environ Pollut 257:113478. https://doi.org/10.1016/j.envpol.2019.113478

Linder G, Lehman CM, Bidwell JR (2010) Ecotoxicology of amphibians and reptiles in a nutshell. In: Sparling DW, Linder G, Bishop CA, Krest SK (eds) Ecotoxicology of amphibians and reptiles, second. CRC Press, Taylor & Francis Group, Boca Raton, pp 69–103

Lopes I, Sedlmayr A, Moreira-Santos M et al (2010) European bee-eater (Merops apiaster) populations under arsenic and metal stress: evaluation of exposure at a mining site. Environ Monit Assess 161:237–245. https://doi.org/10.1007/s10661-008-0741-4

Ma W (2011) Lead in mammals. In: Beyer WN, Meador JP (eds) Enviromental contaminants in biota. Interpreting tissue concentraions, 2nd edn. CRC Press, Taylor & Francis Group, pp 595–608

Macklin MG, Brewer PA, Balteanu D et al (2003) The long term fate and environmental significance of contaminant metals released by the January and March 2000 mining tailings dam failures in Maramures County, upper Tisa Basin, Romania. Appl Geochemistry 18:241–257. https://doi.org/10.1016/S0883-2927(02)00123-3

Madejón P, Domínguez MT, Murillo JM (2009) Evaluation of pastures for horses grazing on soils polluted by trace elements. Ecotoxicology 18:417–428. https://doi.org/10.1007/s10646-009-0296-3

Marques CC, Sánchez-Chardi A, Gabriel SI et al (2007) How does the greater white-toothed shrew, Crocidura russula, responds to long-term heavy metal contamination? – a case study. Sci Total Environ 376:128–133. https://doi.org/10.1016/j.scitotenv.2007.01.061

Marques CC, Gabriel SI, Pinheiro T et al (2008) Metallothionein levels in Algerian mice (Mus spretus) exposed to elemental pollution: an ecophysiological approach. Chemosphere 71:1340–1347. https://doi.org/10.1016/j.chemosphere.2007.11.024

Márquez-Ferrando R, Santos X, Pleguezuelos JM, Ontiveros D (2009) Bioaccumulation of heavy metals in the lizard Psammodromus algirus after a tailing-dam collapse in Aznalcóllar (Southwest Spain). Arch Environ Contam Toxicol 56:276–285. https://doi.org/10.1007/s00244-008-9189-3

Martinez-Haro M, Taggart MA, Lefranc H et al (2013) Monitoring of Pb exposure in waterfowl ten years after a mine spill through the use of noninvasive sampling. PLoS One 8(2):e57295. https://doi.org/10.1371/journal.pone.0057295

Martínez-López E, Gómez-Ramírez P, Espín S et al (2017) Influence of a former mining area in the heavy metals concentrations in blood of free-living mediterranean pond turtles (Mauremys leprosa). Bull Environ Contam Toxicol 99:167–172. https://doi.org/10.1007/s00128-017-2122-6

Mateo R, Taggart MA, Green AJ et al (2006) Altered porphyrin excretion and histopathology of greylag geese (Anser anser) exposed to soil contaminated with lead and arsenic in the Guadalquivir Marshes, southwestern Spain. Environ Toxicol Chem 25:203–212. https://doi.org/10.1897/04-460R.1

Mateo R, Lacorte S, Taggart MA (2016) An overview of recent trends in wildlife ecotoxicology. In: Current trends in wildlife research

Mateos S, Daza P, Domínguez I et al (2008) Genotoxicity detected in wild mice living in a highly polluted wetland area in south western Spain. Environ Pollut 153:590–593. https://doi.org/10.1016/j.envpol.2007.09.008

McIntyre T, Whiting MJ (2012) Increased metal concentrations in giant sungazer lizards (Smaug giganteus) from mining areas in South Africa. Arch Environ Contam Toxicol 63:574–585. https://doi.org/10.1007/s00244-012-9795-y

Meharg AA, Pain DJ, Ellam RM et al (2002) Isotopic identification of the sources of lead contamination for white storks (Ciconia ciconia) in a marshland ecosystem (Doñana, S.W. Spain). Sci Total Environ 300:81–86. https://doi.org/10.1016/S0048-9697(02)00283-8

Méndez-Rodríguez LC, Alvarez-Castañeda ST (2016) Assessment of trace metals in soil, vegetation and rodents in relation to metal mining activities in an arid environment. Bull Environ Contam Toxicol 97:44–49. https://doi.org/10.1007/s00128-016-1826-3

Méndez-Rodríguez L, Álvarez-Castañeda ST (2019) Differences in metal content in liver of heteromyids from deposits with and without previous mining operations. Therya 10:235–242. https://doi.org/10.12933/therya-19-884

Menzie WD, Soto-Viruet Y, Bermudez-Lugo O, et al (2013) Review of selected global mineral industries in 2011 and an outlook to 2017

Millán J, Mateo R, Taggart MA et al (2008) Levels of heavy metals and metalloids in critically endangered Iberian lynx and other wild carnivores from Southern Spain. Sci Total Environ 399:193–201. https://doi.org/10.1016/j.scitotenv.2008.03.038

Milton A, Johnson M (1999) Arsenic in the food chains of a revegetated metalliferous mine tailings pond. Chemosphere 39:765–779

Milton A, Johnson MS, Cook JA (2002) Lead within ecosystems on metalliferous mine tailings in Wales and Ireland. Sci Total Environ 299:177–190

Milton A, Cooke JA, Johnson MS (2003) Accumulation of lead, zinc, and cadmium in a wild population of clethrionomys glareolus from an abandoned lead mine. Arch Environ Contam Toxicol 44:405–411. https://doi.org/10.1007/s00244-002-2014-5

Milton A, Cooke JA, Johnson MS (2004) A comparison of cadmium in ecosystems on metalliferous mine tailings in Wales and Ireland. Water Air Soil Pollut 153:157–172. https://doi.org/10.1023/B:WATE.0000019940.76065.21

Monzalvo-Santos K, Alfaro-De la Torre MC, Chapa-Vargas L et al (2016) Arsenic and lead contamination in soil and in feathers of three resident passerine species in a semi-arid mining region of the Mexican plateau. J Environ Sci Heal – Part A Toxic/Hazardous Subst Environ Eng 51:825–832. https://doi.org/10.1080/10934529.2016.1181451

Moriarty MM, Koch I, Reimer KJ (2012) Arsenic speciation, distribution, and bioaccessibility in shrews and their food. Arch Environ Contam Toxicol 62:529–538. https://doi.org/10.1007/s00244-011-9715-6

Moriarty MM, Koch I, Reimer KJ (2013) Arsenic species and uptake in amphibians (Rana clamitans and Bufo americanus). Environ Sci Process Impacts 15:1520–1528. https://doi.org/10.1039/c3em00223c

Mussali-Galante P, Tovar-Sánchez E, Valverde M et al (2013a) Evidence of population genetic effects in Peromyscus melanophrys chronically exposed to mine tailings in Morelos, Mexico. Environ Sci Pollut Res Int 20:7666–7679. https://doi.org/10.1007/s11356-012-1263-8

Mussali-Galante P, Tovar-Sánchez E, Valverde M, Rojas del Castillo E (2013b) Biomarkers of exposure for assessing environmental metal pollution: from molecules to ecosystems. Rev Int Contam Ambient 29:117–140

Mwase M, Almli B, Sivertsen T et al (2002) Hepatic and renal concentrations of copper and other trace elements in hippopotami (Hippopotamus amphibius L) living in and adjacent to the Kafue and Luangwa Rivers in Zambia. Onderstepoort J Vet Res 69:207–214

Nakayama SMM, Ikenaka Y, Hamada K et al (2011) Metal and metalloid contamination in roadside soil and wild rats around a Pb-Zn mine in Kabwe, Zambia. Environ Pollut 159:175–181. https://doi.org/10.1016/j.envpol.2010.09.007

Nakayama SMM, Ikenaka Y, Hamada K et al (2013) Accumulation and biological effects of metals in wild rats in mining areas of Zambia. Environ Monit Assess 185:4907–4918. https://doi.org/10.1007/s10661-012-2912-6

Niethammer KR, Atkinson RD, Baskett TS, Samson FB (1985) Metals in riparian Wildlife of the Lead Mining District of Southeastern Missouri. Arch Environ Contam Toxicol 14:213–223

Nocete F, Álex E, Nieto JM et al (2005) An archaeological approach to regional environmental pollution in the south-western Iberian Peninsula related to Third millennium BC mining and metallurgy. J Archaeol Sci 32:1566–1576. https://doi.org/10.1016/j.jas.2005.04.012

Nordberg GF, Fowler BA, Nordberg M, Friberg L (2007) Handbook on the toxicology of metals, 3rd edn. Elsevier

Nriagu JO, Pacyna JM (1988) Quantitative assessment of world wide contamination of air, water and soils by trace metals. Nature 333:134–139

Nunes AC, Auffray JC, Mathias ML (2001a) Developmental instability in a riparian population of the Algerian mouse (Mus spretus) associated with a heavy metal-polluted area in central Portugal. Arch Environ Contam Toxicol 41:515–521. https://doi.org/10.1007/s002440010279

Nunes AC, Mathias ML, Crespo AM (2001b) Morphological and haematological parameters in the Algerian mouse (Mus spretus) inhabiting an area contaminated with heavy metals. Environ Pollut 113:87–93

O'Brien EL, Dawson RD (2016) Life-history and phenotypic traits of insectivorous songbirds breeding on reclaimed mine land reveal ecological constraints. Sci Total Environ 553:450–457. https://doi.org/10.1016/j.scitotenv.2016.02.146

O'Brien DJ, Kaneene JB, Poppenga RH (1993) The use of mammals as sentinels for in the environment human exposure to toxic contaminants in the environment. Environ Health Perspect 99:351–368

O'Hara TM, George JC, Blake J et al (2003) Investigation of heavy metals in a large mortality event in Caribou of Northern Alaska. ARTIC 56:125–135

Ollson CA, Koch I, Smith P et al (2009) Addressing arsenic bioaccessibility in ecological risk assessment: a novel approach to avoid overestimating risk. Environ Toxicol Chem 28:668–675. https://doi.org/10.1897/08-204.1

Ortiz-Santaliestra ME, Rodríguez A, Pareja-Carrera J et al (2019) Tools for non-invasive sampling of metal accumulation and its effects in Mediterranean pond turtle populations inhabiting mining areas. Chemosphere 231:194–206. https://doi.org/10.1016/j.chemosphere.2019.05.082

Pareja-Carrera J, Mateo R, Rodríguez-Estival J (2014) Lead (Pb) in sheep exposed to mining pollution: implications for animal and human health. Ecotoxicol Environ Saf 108:210–216. https://doi.org/10.1016/j.ecoenv.2014.07.014

Pascoe GA, Blanchet RJ, Linder G (1994) Bioavailability of metals and arsenic to small mammals at a mining waste-contaminated wetland. Arch Environ Contam Toxicol 27:44–50

Pastor N, López-Lázaro M, Tella JL et al (2001) Assessment of genotoxic damage by the comet assay in white storks (Ciconia ciconia) after the Doñana Ecological Disaster. Mutagenesis 16:219–223

Pastor N, Baos R, López-Lázaro M et al (2004) A 4 year follow-up analysis of genotoxic damage in birds of the Doñana area (south west Spain) in the wake of the 1998 mining waste. Mutagenesis 19:61–65. https://doi.org/10.1093/mutage/geg035

Peakall DB, Walker CH (1994) The role of biomarkers in environmental assessment (3). Vertebrates. Ecotoxicology 3:173–179. https://doi.org/10.1007/BF00117082

Peplow D, Edmonds R (2005) The effects of mine waste contamination at multiple levels of biological organization. Ecol Eng 24:101–119. https://doi.org/10.1016/j.ecoleng.2004.12.011

Pereira R, Pereira ML, Ribeiro R, Gonçalves F (2006) Tissues and hair residues and histopathology in wild rats (Rattus rattus L.) and Algerian mice (Mus spretus Lataste) from an abandoned mine area (Southeast Portugal). Environ Pollut 139:561–575. https://doi.org/10.1016/j.envpol.2005.04.038

Phelps KL, McBee K (2009) Ecological characteristics of small mammal communities at a superfund site. Am Midl Nat 161:57–68. https://doi.org/10.1674/0003-0031-161.1.57

Phelps KL, Mcbee K (2010) Population parameters of Peromyscus leucopus (White- Footed Deermice) inhabiting a heavy metal contaminated superfund site. Southwest Nat 55:363–373

Purcell PW, Hynes MJ, Fairley JS (1992) Lead levels in Irish small rodents (Apodemus sylvaticus and Clethrionomys glareolus) from around a Tailings Pond and along Motorway Verges. Proc R Ir Acad B 92B:79–90

Qiu G, Abeysinghe KS, Yang XD et al (2019) Effects of selenium on mercury bioaccumulation in a terrestrial food chain from an abandoned mercury mining region. Bull Environ Contam Toxicol 102:329–334. https://doi.org/10.1007/s00128-019-02542-z

Quina AS, Durão AF, Muñoz-Muñoz F et al (2019) Population effects of heavy metal pollution in wild Algerian mice (Mus spretus). Ecotoxicol Environ Saf 171:414–424. https://doi.org/10.1016/j.ecoenv.2018.12.062

Rattner BA (2009) History of wildlife toxicology. Ecotoxicology 18:773–783. https://doi.org/10.1007/s10646-009-0354-x

Reglero MM, Monsalve-González L, Taggart MA, Mateo R (2008) Transfer of metals to plants and red deer in an old lead mining area in Spain. Sci Total Environ 406:287–297. https://doi.org/10.1016/j.scitotenv.2008.06.001

Reglero MM, Taggart MA, Castellanos P, Mateo R (2009a) Reduced sperm quality in relation to oxidative stress in red deer from a lead mining area. Environ Pollut 157:2209–2215. https://doi.org/10.1016/j.envpol.2009.04.017

Reglero MM, Taggart MA, Monsalve-González L, Mateo R (2009b) Heavy metal exposure in large game from a lead mining area: effects on oxidative stress and fatty acid composition in liver. Environ Pollut 157:1388–1395. https://doi.org/10.1016/j.envpol.2008.11.036

Reichl C, Schatz M (2019) World mining data 2019. Vienna

Rico M, Benito G, Salgueiro AR et al (2008) Reported tailings dam failures. A review of the European incidents in the worldwide context. J Hazard Mater 152:846–852. https://doi.org/10.1016/j.jhazmat.2007.07.050

Roberts RD, Johnson MS (1978) Dispersal of heavy metals from abandoned mine workings and their transference through terrestrial food chains. Environ Pollut 16:293–310

Roberts RD, Johnson MS, Hutton M (1978) Lead contamination of small mammals from abandoned metalliferous mines. Environ Pollut 15:61–69. https://doi.org/10.1016/0013-9327(78)90061-7

Rodríguez Álvarez C, Jiménez Moreno M, López Alonso L et al (2013) Mercury, methylmercury, and selenium in blood of bird species from Doñana National Park (Southwestern Spain) after a mining accident. Environ Sci Pollut Res Int 20:5361–5372. https://doi.org/10.1007/s11356-013-1540-1

Rodríguez-Estival J, Mateo R (2019) Exposure to anthropogenic chemicals in wild carnivores: a silent conservation threat demanding long-term surveillance. Curr Opin Environ Sci Heal 11:21–25. https://doi.org/10.1016/j.coesh.2019.06.002

Rodríguez-Estival J, Martinez-Haro M, Monsalve-González L, Mateo R (2011a) Interactions between endogenous and dietary antioxidants against Pb-induced oxidative stress in wild ungulates from a Pb polluted mining area. Sci Total Environ 409:2725–2733. https://doi.org/10.1016/j.scitotenv.2011.04.010

Rodríguez-Estival J, Taggart MA, Mateo R (2011b) Alterations in vitamin A and E levels in liver and testis of wild ungulates from a lead mining area. Arch Environ Contam Toxicol 60:361–371. https://doi.org/10.1007/s00244-010-9597-z

Rodríguez-Estival J, Barasona JA, Mateo R (2012) Blood Pb and δ-ALAD inhibition in cattle and sheep from a Pb-polluted mining area. Environ Pollut 160:118–124. https://doi.org/10.1016/j.envpol.2011.09.031

Rodríguez-Estival J, Álvarez-Lloret P, Rodríguez-Navarro AB, Mateo R (2013a) Chronic effects of lead (Pb) on bone properties in red deer and wild boar: relationship with vitamins A and D3. Environ Pollut 174:142–149. https://doi.org/10.1016/j.envpol.2012.11.019

Rodríguez-Estival J, de la Lastra JMP, Ortiz-Santaliestra ME et al (2013b) Expression of immunoregulatory genes and its relationship to lead exposure and lead-mediated oxidative stress in wild ungulates from an abandoned mining area. Environ Toxicol Chem 32:876–883. https://doi.org/10.1002/etc.2134

Rodríguez-Estival J, Sánchez MI, Ramo C et al (2019) Exposure of black-necked grebes (Podiceps nigricollis) to metal pollution during the moulting period in the Odiel Marshes, Southwest Spain. Chemosphere 216:774–784. https://doi.org/10.1016/j.chemosphere.2018.10.145

Rodríguez-Estival J, Ortiz-Santaliestra ME, Mateo R (2020) Assessment of ecotoxicological risks to river otters from ingestion of invasive red swamp crayfish in metal contaminated areas: use of feces to estimate dietary exposure. Environ Res 181:108907. https://doi.org/10.1016/j.envres.2019.108907

Ruiz-Laguna J, García-Alfonso C, Peinado J et al (2001) Biochemical biomarkers of pollution in Algerian mouse (Mus spretus) to assess the effects of The Aznalcóllar disaster on Doñana Park (Spain). Biomarkers 6:146–160

Sánchez-Chardi A (2007) Tissue, age, and sex distribution of thallium in shrews from Doñana, a protected area in SW Spain. Sci Total Environ 383:237–240. https://doi.org/10.1016/j.scitotenv.2007.05.017

Sánchez-Chardi A, Nadal J (2007) Bioaccumulation of metals and effects of landfill pollution in small mammals. Part I. The greater white-toothed shrew, Crocidura russula. Chemosphere 68:703–711. https://doi.org/10.1016/j.chemosphere.2007.01.042

Sánchez-Chardi A, López-Fuster MJ, Nadal J (2007a) Bioaccumulation of lead, mercury, and cadmium in the greater white-toothed shrew, Crocidura russula, from the Ebro Delta (NE Spain): sex- and age-dependent variation. Environ Pollut 145:7–14. https://doi.org/10.1016/j.envpol.2006.02.033

Sánchez-Chardi A, Marques CC, Nadal J, da Luz MM (2007b) Metal bioaccumulation in the greater white-toothed shrew, Crocidura russula, inhabiting an abandoned pyrite mine site. Chemosphere 67:121–130. https://doi.org/10.1016/j.chemosphere.2006.09.009

Sánchez-Chardi A, Peñarroja-Matutano C, Oliveira Ribeiro CA, Nadal J (2007c) Bioaccumulation of metals and effects of a landfill in small mammals. Part II. The wood mouse, Apodemus sylvaticus. Chemosphere 70:101–109. https://doi.org/10.1016/j.chemosphere.2007.06.047

Sánchez-Chardi A, Marques CC, Gabriel SI et al (2008) Haematology, genotoxicity, enzymatic activity and histopathology as biomarkers of metal pollution in the shrew Crocidura russula. Environ Pollut 156:1332–1339. https://doi.org/10.1016/j.envpol.2008.02.026

Sánchez-Chardi A, Oliveira Ribeiro CA, Nadal J (2009) Metals in liver and kidneys and the effects of chronic exposure to pyrite mine pollution in the shrew Crocidura russula inhabiting the protected wetland of Doñana. Chemosphere 76:387–394. https://doi.org/10.1016/j.chemosphere.2009.03.036

Sánchez-Virosta P, Espín S, García-Fernández AJ, Eeva T (2015) A review on exposure and effects of arsenic in passerine birds. Sci Total Environ 512–513:506–525. https://doi.org/10.1016/j.scitotenv.2015.01.069

Satta A, Verdinelli M, Ruiu L et al (2012) Combination of beehive matrices analysis and ant biodiversity to study heavy metal pollution impact in a post-mining area (Sardinia, Italy). Environ Sci Pollut Res Int 19:3977–3988. https://doi.org/10.1007/s11356-012-0921-1

Sault N (2018) Condors, water, and mining: heeding voices from Andean communities. Ethnobiol Lett 9:27–43. https://doi.org/10.14237/ebl.9.1.2018.1079

Saunders JR, Knopper LD, Yagminas A et al (2009) Use of biomarkers to show sub-cellular effects in meadow voles (Microtus pennsylvanicus) living on an abandoned gold mine site. Sci Total Environ 407:5548–5554. https://doi.org/10.1016/j.scitotenv.2009.07.026

Saunders JR, Knopper LD, Koch I, Reimer KJ (2010) Arsenic transformations and biomarkers in meadow voles (Microtus pennsylvanicus) living on an abandoned gold mine site in Montague, Nova Scotia, Canada. Sci Total Environ 408:829–835. https://doi.org/10.1016/j.scitotenv.2009.11.006

Saunders JR, Hough C, Knopper LD et al (2011) Arsenic transformations in terrestrial small mammal food chains from contaminated sites in Canada. J Environ Monit 13:1784. https://doi.org/10.1039/c1em10225g

Shahsavari A, Tabatabaei Yazdi F, Moosavi Z et al (2019) A study on the concentration of heavy metals and histopathological changes in Persian jirds (Mammals; Rodentia), affected by mining activities in an iron ore mine in Iran. Environ Sci Pollut Res:12590–12604. https://doi.org/10.1007/s11356-019-04646-9

Sharma RK, Agrawal M (2005) Biological effects of heavy metals: an overview. J Environ Biol 26:301–313

Sierra-Marquez L, Peñuela-Gomez S, Franco-Espinosa L et al (2018) Mercury levels in birds and small rodents from Las Orquideas National Natural Park, Colombia. Environ Sci Pollut Res 25:35055–35063. https://doi.org/10.1007/s11356-018-3359-2

Sileo L, Creekmore LH, Audet DJ et al (2001) Lead poisoning of waterfowl by contaminated sediment in the Coeur d'Alene River. Arch Environ Contam Toxicol 41:364–368. https://doi.org/10.1007/s002440010260

Sileo L, Beyer WN, Mateo R (2003) Pancreatitis in wild zinc-poisoned waterfowl. Avian Pathol 32:655–660. https://doi.org/10.1080/03079450310001636246

Smith GJ, Rongstad OJ (1982) Small mammal heavy metal concentrations from mined and control sites. Environ Pollut 28:121–134. https://doi.org/10.1016/0143-1471(82)90098-8

Smith PN, Cobb GP, Godard-Codding C et al (2007) Contaminant exposure in terrestrial vertebrates. Environ Pollut 150:41–64. https://doi.org/10.1016/j.envpol.2007.06.009

Smith KM, Dagleish MP, Abrahams PW (2010) The intake of lead and associated metals by sheep grazing mining-contaminated floodplain pastures in mid-Wales, UK: II. Metal concentrations in blood and wool. Sci Total Environ 408:1035–1042. https://doi.org/10.1016/j.scitotenv.2009.10.023

Smits JEG, Bortolotti GR, Baos R et al (2005) Skeletal pathology in white storks (Ciconia ciconia) associated with heavy metal contamination in southwestern Spain. Toxicol Pathol 33:441–448. https://doi.org/10.1080/01926230590953097

Smits JE, Bortolotti GR, Baos R et al (2007) Disrupted bone metabolism in contaminant-exposed white storks (Ciconia ciconia) in southwestern Spain. Environ Pollut 145:538–544. https://doi.org/10.1016/j.envpol.2006.04.032

Strom SM, Ramsdell HS, Archuleta AS (2002) Aminolevulinic acid dehydratase activity in American dippers (Cinclus mexicanus) from a metal-impacted stream. Environ Toxicol Chem 21:115–120. https://doi.org/10.1897/1551-5028(2002)021<0115:aadaia>2.0.co;2

Syakalima MS, Choongo KC, Chilonda P et al (2001a) Bioaccumulation of lead in wildlife dependent on the contaminated environment of the Kafue Flats. Bull Environ Contam Toxicol 67:438–445. https://doi.org/10.1007/s00128-001-0143-6

Syakalima MS, Choongo KC, Nakazato Y et al (2001b) An investigation of heavy metal exposure and risks to wildlife in the Kafue flats of Zambia. J Vet Med Sci 63:315–318. https://doi.org/10.1292/jvms.63.315

Taggart MA, Figuerola J, Green AJ et al (2006) After the Aznalcóllar mine spill: arsenic, zinc, selenium, lead and copper levels in the livers and bones of five waterfowl species. Environ Res 100:349–361. https://doi.org/10.1016/j.envres.2005.07.009

Taggart MA, Reglero MM, Camarero PR, Mateo R (2011) Should legislation regarding maximum Pb and Cd levels in human food also cover large game meat? Environ Int 37:18–25. https://doi.org/10.1016/j.envint.2010.06.007

Tanzarella C, Degrassi F, Cristaldi M et al (2001) Genotoxic damage in free-living Algerian mouse (Mus spretus) after the Coto Doñana ecological disaster. Environ Pollut 115:43–48. https://doi.org/10.1016/S0269-7491(01)00092-6

Tchounwou PB, Yedjou CG, Patlolla AK, Sutton DJ (2012) Heavy metal toxicity and the environment. In: Molecular, clinical and environmental toxicology, pp 133–164

Tovar-Sánchez E, Cervantes LT, Martínez C et al (2012) Comparison of two wild rodent species as sentinels of environmental contamination by mine tailings. Environ Sci Pollut Res Int 19:1677–1686. https://doi.org/10.1007/s11356-011-0680-4

Tyler Miller GJ, Spoolman SE (2009) Geology and nonrenewable minerals. In: Living in the environment: concepts, connections, and solutions, 16th edn. Brooks/Cole, Belmont

Udroiu I, Cristaldi M, Ieradi LA et al (2008) Biomonitoring of Doñana National Park using the Algerian mouse (Mus spretus) as a sentinel species. Fresen Environ Bull 17:1519–1525

UNDP, UN Environment (2018) Managing mining for sustainable development: a sourcebook. United Nat, Bangkok

UNEP (2000) Mining and sustainable development II: challenges and perspectives. United Nations Environment Programme Division of Technology Industry and Economics. Ind Environ 23:1–96

Valladares FP, Alvarado S, Urra RC et al (2013) Cadmium and lead content in liver and kidney tissues of wild turkey vulture Cathartes aura (Linneo, 1758) from Chañaral, Atacama desert, Chile. Gayana (Concepción) 77:97–104. https://doi.org/10.4067/s0717-65382013000200004

van den Brink NW, Lammertsma DR, Dimmers WJ, Boerwinkel MC (2011) Cadmium accumulation in small mammals: species traits, soil properties, and spatial habitat use. Environ Sci Technol 45:7497–7502. https://doi.org/10.1021/es200872p

van der Merwe D, Carpenter JW, Nietfeld JC, Miesner JF (2011) Adverse health effects in Canada geese (Branta canadensis) associated with waste from zinc and lead mines in the Tri-State Mining District (Kansas, Oklahoma, and Missouri, USA). J Wildl Dis 47:650–660. https://doi.org/10.7589/0090-3558-47.3.650

van Ooik T, Pausio S, Rantala MJ (2008) Direct effects of heavy metal pollution on the immune function of a geometrid moth, Epirrita autumnata. Chemosphere 71:1840–1844. https://doi.org/10.1016/j.chemosphere.2008.02.014

Vanparys C, Dauwe T, Van Campenhout K et al (2008) Metallothioneins (MTs) and delta-aminolevulinic acid dehydratase (ALAd) as biomarkers of metal pollution in great tits (Parus major) along a pollution gradient. Sci Total Environ 401:184–193. https://doi.org/10.1016/j.scitotenv.2008.04.009

Venkateswarlu K, Nirola R, Kuppusamy S et al (2016) Abandoned metalliferous mines: ecological impacts and potential approaches for reclamation. Rev Environ Sci Biotechnol 15:327–354. https://doi.org/10.1007/s11157-016-9398-6

Viegas-Crespo AM, Lopes PA, Pinheiro MT et al (2003) Hepatic elemental contents and antioxidant enzyme activities in Algerian mice (Mus spretus) inhabiting a mine area in central Portugal. Sci Total Environ 311:101–109. https://doi.org/10.1016/S0048-9697(03)00136-0

Weech SA, Scheuhammer AM, Elliott JE (2006) Mercury exposure and reproduction in fish-eating birds breeding in the Pinchi Lake region, British Columbia, Canada. Environ Toxicol Chem 25:1433–1440. https://doi.org/10.1897/05-181R.1

Wilson B, Pyatt FB (2007) Heavy metal dispersion, persistance, and bioccumulation around an ancient copper mine situated in Anglesey, UK. Ecotoxicol Environ Saf 66:224–231. https://doi.org/10.1016/j.ecoenv.2006.02.015

Yabe J, Nakayama SMM, Ikenaka Y et al (2011) Uptake of lead, cadmium, and other metals in the liver and kidneys of cattle near a lead-zinc mine in Kabwe, Zambia. Environ Toxicol Chem 30:1892–1897. https://doi.org/10.1002/etc.580

Yang F, Xie S, Liu J et al (2018) Arsenic concentrations and speciation in wild birds from an abandoned realgar mine in China. Chemosphere 193:777–784. https://doi.org/10.1016/j.chemosphere.2017.11.098

Younger PL, Banwart SA, Hedin RS (2002) Mine water: hydrology, pollution, remediation. Kluwer Academic Press, Dordrecht

Fluorotelomer Alcohols' Toxicology Correlates with Oxidative Stress and Metabolism

Yujuan Yang, Kuiyu Meng, Min Chen, Shuyu Xie, and Dongmei Chen

Contents

1 Introduction 73
2 Metabolism of FTOHs 77
   2.1 Metabolic Pathways and Toxicokinetics 77
   2.2 Metabolising Enzymes 80
3 Toxicity 81
4 Toxicity Induced by Oxidative Stress 84
   4.1 Generation of Oxidative Stress and ROS 84
   4.2 Oxidative Damage 89
   4.3 Alterations in Antioxidant Status 90
   4.4 Cell Signalling 91
5 Conclusion and Prospects 92
References 94

Abstract Fluorotelomer alcohols (FTOHs) are widely used as industrial raw materials due to their unique hydrophobic and oleophobic properties. However, because of accidental exposure to products containing FTOHs or with the widespread use of FTOHs, they tend to contaminate the water and the soil. There are reports demonstrating that FTOHs can cause various harmful effects in animals and humans (for example, neurotoxicity, hepatotoxicity, nephrotoxicity, immunotoxicity, endocrine-disrupting activity, and developmental and reproductive toxicities). Oxidative stress

---

Y. Yang · K. Meng · M. Chen · S. Xie
National Reference Laboratory of Veterinary Drug Residues (HZAU) and MAO Key Laboratory for Detection of Veterinary Drug Residues, Wuhan, Hubei, China
e-mail: 2397889344@qq.com; 978648029@qq.com; 2049976158@qq.com; 41098641@qq.com

D. Chen (✉)
National Reference Laboratory of Veterinary Drug Residues (HZAU) and MAO Key Laboratory for Detection of Veterinary Drug Residues, Wuhan, Hubei, China

MOA Laboratory for Risk Assessment of Quality and Safety of Livestock and Poultry Products, Huazhong Agricultural University, Wuhan, Hubei, China
e-mail: dmchen1027@163.com

© The Author(s), under exclusive license to Springer Nature Switzerland AG 2020
P. de Voogt (ed.), *Reviews of Environmental Contamination and Toxicology Volume 256*,
Reviews of Environmental Contamination and Toxicology 256,
https://doi.org/10.1007/398_2020_57

is related to a variety of toxic effects induced by FTOHs. To date, few reviews have addressed the relationship between the toxicity of FTOHs and oxidative stress. This article summarises research demonstrating that the toxicity induced by FTOHs correlates with oxidative stress and metabolism. Furthermore, during the metabolic process of FTOHs, a number of cytochrome P450 enzymes (CYP450) are involved and many metabolites are produced by these enzymes, which can induce oxidative stress. This is also reviewed.

Keywords FTOHs · Metabolism · Oxidative stress · PFOA · ROS · Toxicity

Abbreviations

4-3 FTCA	4-3 Fluorotelomer carboxylic acid
5-3 FTCA	5-3 Fluorotelomer carboxylic acid
6-2 FTAL	6-2 Fluorotelomer aldehyde
6-2 FTCA	6-2 Saturated fluorotelomer carboxylic acids
6-2 FTUCA	6-2 Unsaturated fluorotelomer carboxylic acids
7-2 sFTOH	7-2 Secondary fluorotelomer alcohol
7-3 AL	7-3 Aldehyde
7-3 FTCA	7-3 Acid
7-3 TA	7-3 Acid taurine
7-3 UAL	7-3 α-β Unsaturated aldehyde
7-3UA	7-3 Unsaturated acid
8-2 FTAL	8-2 Fluorotelomer aldehyde
8-2 FTCA	8-2 Fluorotelomer acid
8-2 FTOH-Sulf	8-2 Fluorotelomer alcohol sulphate
8-2 FTUAL	8-2 Fluorotelomer α, β-unsaturated aldehyde
8-2 FTUCA	8-2 Fluorotelomer unsaturated acid
8-2 uFTOH-GS	8-2 Unsaturated alcohol conjugate
8-OH-dG	8-Hydroxydeoxyguanosine
CAT	Catalase
CYP450	Cytochrome P450 enzymes
EC50	Concentration causing 50% cell death
FCSs	Food contact substances
FTOHs	Fluorotelomer alcohols
GP_X	Glutathione peroxidase
GR	Glutathione reductase
GSH	Glutathione
GST	Glutathione S-transferase
H_2O_2	Hydrogen peroxide
hER	Human estrogen receptor
LC50	Median lethal concentration
LD50	Median lethal dose

MDA	Malondialdehyde
NOAEL	No-observed-adverse-effect level
O_2^-	Superoxide anion radical
PFASs	Polyfluoroalkyl substances
PFBA	Perfluorobutanoic acid
PFCAs	Perfluoroalkyl carboxylic acids
PFCs	Poly- and perfluorinated compounds
PFHpA	Perfluoroheptanoic acid
PFHxA	Perfluorohexanoic acid
PFNA	Perfluorononanoic acid
PFOA	Perfluorooctanoic acid
PFOS	Perfluorooctanesulfonic acid
PFOSA	Perfluorooctanesulfonamide
PFPeA	Perfluoropentanoic acid
ROS	Reactive oxygen species
SOD	Superoxide dismutase

1 Introduction

Fluorotelomer alcohols (FTOHs) constitute a unique type of polyfluoroalkyl substances (PFASs). They contain stable fluorinated carbons that give them thermal and chemical stability. They are industrial raw materials composed of linear alkyl groups and terminal ethanol ($F(CF_2)xC_2H_4OH$; x = 6, 8 and 10). Due to their hydrophobicity and oleophobicity, FTOHs have become basic materials that are used to manufacture industrial and consumer products (Buck et al. 2011; Fasano et al. 2006). In commercial use, fluorotelomer alcohols mainly consist of three groups according their chain length: 6-2, 8-2 and 10-2 FTOH (Kotthoff et al. 2015; Ritter 2015). Polymers made from FTOHs have been approved for widespread use in food contact packages such as in paper, boxes, bags and paperboard to repel oil and grease (Rice 2015). An estimated 20 million pounds of FTOHs were produced in 2006 (Huang et al. 2019). Research has reported that FTOHs have been detected in Asia, Europe, the US and Africa. They have also been found in waste water, air, airborne particles and settled dust (Fraser et al. 2013; Heydebreck et al. 2016; Lai et al. 2016; Sha et al. 2018; Shoeib et al. 2016; Wang et al. 2018; Winkens et al. 2018; Wong et al. 2018; Zhao et al. 2017b). The FTOH air concentration is higher than other PFASs, and the geometric mean of 8-2 FTOH was reportedly 9,920 pg/m^3 (Fraser et al. 2012). In the US, the concentrations of 6-2 FTOH detected in eco-friendly paper tableware, food contact paper, waxes and textiles were 485–499 ng/g, 12,700 ng/g, 331,000 ng/g and 40,900 ng/g, respectively (Liu et al. 2015b; Yuan et al. 2016). In surface waters, PFOS levels were determined as 60 ng/l, 88 ng/l and 5.5 ng/l in the US, Canada and China (Zareitalabad et al. 2013). The PFOA concentrations in South Africa were 310 and 1,089 ng/l (Ssebugere et al. 2020).

In some cases, it was higher in drinking water than the Environmental Protection Agency's safe levels (70 ng/l). PFOS concentrations in occupational exposure areas ranged from 10.49 μg/g to 4691.94 μg/g (dust) (Wang et al. 2010). This widespread distribution creates more possibilities for human exposure through multiple inhalation and ingestion methods.

Due to the extensive global application of FTOHs, humans may be inadvertently exposed to these compounds. In the US population, the concentrations of PFOS, PFOA, PFHxA and 8-2 FTOH-sulphate detected in the serum were 8.9 ng/ml, 3.28 ng/ml, 1.8 ng/mL and 0.08 ng/mL, respectively (Dagnino et al. 2016). The serum levels of PFOA in humans show a significant correlation with airborne FTOH levels (Fraser et al. 2012). Office worker, ski waxers, manufacturers, outdoor equipment vendors and residents and workers in renovated buildings may be exposed to FTOHs.

FTOHs can cause various toxicities to humans and wildlife because of their accumulation and persistence in the environment and mammalian tissues, including neurotoxicity, hepatotoxicity, nephrotoxicity, immunotoxicity, endocrine-disrupting activity, developmental and reproductive toxicities (Nilsson et al. 2010a, b, 2013; Perkins et al. 2004). Thus, it is essential for research to focus more on the toxicity of FTOHs. It is probable that the toxicity of FTOHs is mainly related to their metabolites. To date, several reviews have focused on research into the biotransformation of FTOHs in animals and aquatic environments, among others (Cui et al. 2016; Butt et al. 2014; Butt et al. 2010a; Froemel and Knepper 2010; Hekster et al. 2003). In typical biotransformation processes in microbial communities and biological systems, FTOHs can degrade to a variety of PFCAs (Russell et al. 2015; Wang et al. 2005b; Zhang et al. 2013b, 2017a, b).

Many articles have been published associating oxidative stress and FTOHs' metabolites with the various toxicities of FTOHs (Rainieri et al. 2017; Martin et al. 2009). An increasing amount of evidence has suggested that oxidative stress may play a role in leading to the toxic effects of FTOHs. Therefore, it is necessary to review the progress regarding the toxicity and possible toxic mechanisms of FTOHs.

This study analysed and summarised the literature over the past decade that correlates with oxidative stress and the underlying mechanism of toxicities induced by FTOHs and their metabolites. This includes the damage, alteration of antioxidant status and signalling pathways associated with oxidative stress mediated by FTOHs. The metabolic pathways and toxicokinetics during FTOH metabolism are also discussed. The various CYP450 enzymes involved are also reviewed. This review proposes a new idea on FTOHs' toxicity mechanisms and preventing human and animal poisoning. The structures of FTOHs and their metabolites are shown in Table 1.

Table 1 Structures of FTOHs and its metabolites

Compound name	Molecular formula	Compound structure
8-2 FTOH	$CF_3(CF_2)_7CH_2CH_2OH$	
8-2 FTAL	$CF_3(CF_2)_7CH_2CHO$	
8-2 FTUAL	$CF_3(CF_2)_6CF=CHCHO$	
8-2 FTUAL-GS	$CF_3(CF_2)_6C(SG)=CHCHO$	
8-2 uFTOH-GS	$CF_3(CF_2)_6C(SG)=CHCH_2OH$	
8-2 FTCA	$CF_3(CF_2)_7CH_2COOH$	
8-2 FTUCA	$CF_3(CF_2)_6CF=CHCOOH$	
7-3 UA	$CF_3(CF_2)_6CH=CHCOOH$	
7-3 unsaturated amide	$CF_3(CF_2)_6CH=CHCONH_2$	
7-3 FTCA	$CF_3(CF_2)_6CH_2CH_2COOH$	
7-3 TA	$CF_3(CF_2)_6CH_2CH_2CONHC_2H_5SO_3$	
7-3 UAL	$F(CF_2)_7CH=CHCHO$	
7-3 AL	$CF_3(CF_2)_6C_2H_4CHO$	

(continued)

Table 1 (continued)

Compound name	Molecular formula	Compound structure
PFHpA	F(CF$_2$)$_6$COOH	
PFOA	F(CF$_2$)$_7$COOH	
7-3β-keto aldehyde	F(CF$_2$)$_7$COCH$_2$CHO	
7-3 β-keto acid	F(CF$_2$)$_7$COCH$_2$COOH	
7-2 ketone	F(CF$_2$)$_7$COCH$_3$	
7-2 sFTOH	F(CF$_2$)$_7$CH(OH)CH$_3$	
PFNA	F(CF$_2$)$_7$CF$_2$COOH	
PFPeA	F(CF$_2$)$_4$COOH	
PFHxA	F(CF$_2$)$_5$COOH	
6-2 FTOH	CF$_3$(CF$_2$)$_5$CH$_2$CH$_2$OH	
6-2 FTAL	CF$_3$(CF$_2$)$_5$CH$_2$CHO	
6-2 FTCA	CF$_3$(CF$_2$)$_5$CH$_2$COOH	
6-2 FTUCA	CF$_3$(CF$_2$)$_4$CF=CHCOOH	

(continued)

Table 1 (continued)

Compound name	Molecular formula	Compound structure
5-3FTUCA	CF$_3$(CF$_2$)$_4$CH=CHCOOH	
5-2 ketone	F(CF$_2$)$_5$COCH$_3$	
5-2 sFTOH	F(CF$_2$)$_5$CH(OH)CH$_3$	
5-3 FTCA	CF$_3$(CF$_2$)$_4$CH$_2$CH$_2$COOH	
4-3 FTCA	CF$_3$(CF$_2$)$_3$CH$_2$CH$_2$COOH	
5-3 unsaturated amide	CF$_3$(CF$_2$)$_4$CH=CHCONH$_2$	
PFBA	F(CF$_2$)$_3$COOH	

Note: The reference in Table 1 was cited from Wang et al. (2005a), Martin et al. (2005, 2009), Nabb et al. (2007), Wang et al. (2009), Fasano et al. (2009), Ruan et al. (2014) and Russell et al. (2015)

2 Metabolism of FTOHs

2.1 Metabolic Pathways and Toxicokinetics

It is probable that the toxicity of FTOHs is mainly related to their metabolites. It has been reported that FTOHs are classic precursors that can metabolise to various products including PFCAs, fluorotelomer aldehyde and conjugates (Butt et al. 2014; Dinglasan et al. 2004; Ellis et al. 2004; Wang et al. 2005a, b).

8-2 FTOH can ultimately metabolise into perfluorocarboxylic acids, such as perfluoropentanoic acid (PFPeA), perfluorohexanoic acid (PFHxA), perfluoroheptanoic acid (PFHpA), PFOA and PFNA. The mid-metabolites (FTUAL, 7-2 sFTOH) of 8-2 FTOH can bind with proteins and cause toxicity, and the final metabolite, PFOA, is an established peroxisome proliferator. 8-2 FTOH is metabolised through two major pathways including phase I and II metabolism. The

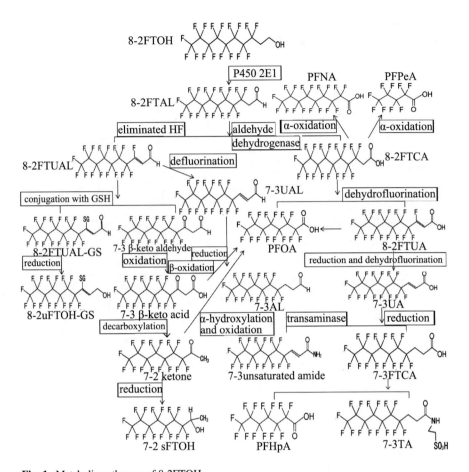

Fig. 1 Metabolic pathways of 8-2FTOH

assumed metabolic pathway of 8-2 FTOH is shown in Fig. 1, which integrates the previous related research. In phase II metabolism, 8-2 FTOH can conjugate with sulphate and glucuronide. In phase I metabolism, 8-2 FTOH first is oxidised by P450 2E1 to form a transient intermediate, 8-2 fluorotelomer aldehyde (8-2 FTAL), and then 8-2 FTAL yields 8-2 FTUAL by eliminating HF without enzyme catalysis. 8-2 FTAL can also be oxidised by aldehyde dehydrogenase to form 8-2 fluorotelomer acid (8-2 FTCA), and then 8-2 FTUCA is formed by dehydrofluorination of 8-2 FTCA. 8-2 FTUCA was reduced and dehydrofluorinated to yield 7-3 unsaturated acid (7-3UA) and then reduced to form 7-3 acid (7-3 FTCA), ultimately leading to conjugation with taurine yielding 7-3 acid taurine (7-3 TA) or forming PFHpA. Furthermore, 7-3 UA may also be catalysed by a transaminase yielding 7-3 unsaturated amide (Wang et al. 2005a).

Otherwise, 8-2 FTUAL was conjugated with GSH to form 8-2 FTUAL-GS, and then underwent reduction to form the unsaturated alcohol conjugate (8-2 uFTOH-

GS), which proceeded by defluorination to 7-3 α-β-unsaturated aldehyde (7-3 UAL) and reduction to 7-3 aldehyde (7-3 AL) (Nabb et al. 2007). It is proposed that 7-3 β-keto aldehyde forms from 8-2 FTUAL biotransformation and can be oxidised to form 7-3 β-keto acid. After 7-2 ketone forms through 7-3 β-keto acid, it can not only be α-hydroxylated and oxidised to yield the final metabolites PFOA but can also be reduced to 7-2 secondary fluorotelomer alcohol (7-2 sFTOH). Alternatively, PFOA was proposed to be obtained after 7-3 β-keto acid was β-oxidised. Martin et al. reported that PFOA is probably produced through β-oxidation and hypothesised that 8-2 FTCA and 8-2 FTUCA may be catalysed by the thiolase enzyme to form PFOA-S-CoA β-ketoacyl substrate, which is processed similar to the progression of the β-oxidation cycle (Martin et al. 2005). However, Wang et al. argued that 8-2 FTCA and 8-2 FTUCA would not be β-oxidised because H atoms are not present on the β-carbon of 8-2 FTCA and only one H atom is present in the β-carbon of 8-2 FTUCA. Therefore, it has been hypothesised that PFOA forms through β-oxidation of 7-3 acid (Wang et al. 2005a, b). However, it has been shown that there was no PFOA after 7-3 acid or 7-3 UA were incubated with rat, human or trout hepatocyte preparations, which demonstrates that β-oxidation from 7-3 acid or 7-3 UA is not an active metabolic pathway (Butt et al. 2010b; Nabb et al. 2007). PFNA and PFPeA were proposed to be formed from 8-2 FTCA through direct α-oxidation, while PFHpA is formed from 7-3 FTCA.

6-2 FTOH's metabolites mainly include 6-2 FTAL, 6-2 FTCA, 6-2 FTUCA, perfluorobutanoic acid (PFBA), PFPeA, PFHxA, PFHpA, 5-3 FTCA, and 4-3 FTCA (Fig. 2). Kidneys and livers are the primary sites of 6-2 FTOH metabolism in rodents and humans (Gannon et al. 2012; Ruan et al. 2014). First, 6-2 FTOH is converted into transient intermediates such as 6-2 FTAL and 6-2 FTCA; 6-2 FTUCA can then be conjugated by phase II enzymes to yield conjugates (reacting with glutathione, glucuronide, and sulphate) through the main metabolic pathway (Ruan et al. 2014; Russell et al. 2015). 6-2 FTAL, 6-2 FTCA, and 6-2 FTUCA may also undergo phase I metabolism to form PFCAs, including PFBA, PFPeA, PFHxA, PFHpA, 5-3 FTCA and 4-3 FTCA.

Most studies use a one compartment model described in the metabolism of 6-2 FTOH in rats. 6-2 FTOH is a rapidly formed metabolite with a short $T_{1/2} < 3$ days (Zhang et al. 2017a). After rat hepatocytes were treated with 8-2 FTOH for 4 h, most of the parent compound disappeared (78%) (Martin et al. 2005). Kudo et al. (2005) reported that the detection of intermediate metabolites reached a peak in the liver of mice after they were treated for 6 h. A study conducted in foetal and neonatal mice could only detect the final metabolites PFOA and PFNA (Henderson and Smith 2007). The metabolic rate of 8-2 FTOH was faster than 8-2 FTCA and 8-2 FTUCA in hepatocyte incubation, but the molar balance for 8-2 FTOH was much less than 8-2 FTCA and 8-2 FTUCA (Martin et al. 2005).

The elimination rate and intrinsic clearance of 8-2 FTOH in rodents was faster than in humans and trout, and the elimination rate in humans was faster than in trout, but the intrinsic clearance between humans and trout showed no significant differences (Nabb et al. 2007). After rainbow trout were fed with 8-2 FTOH and 10-2

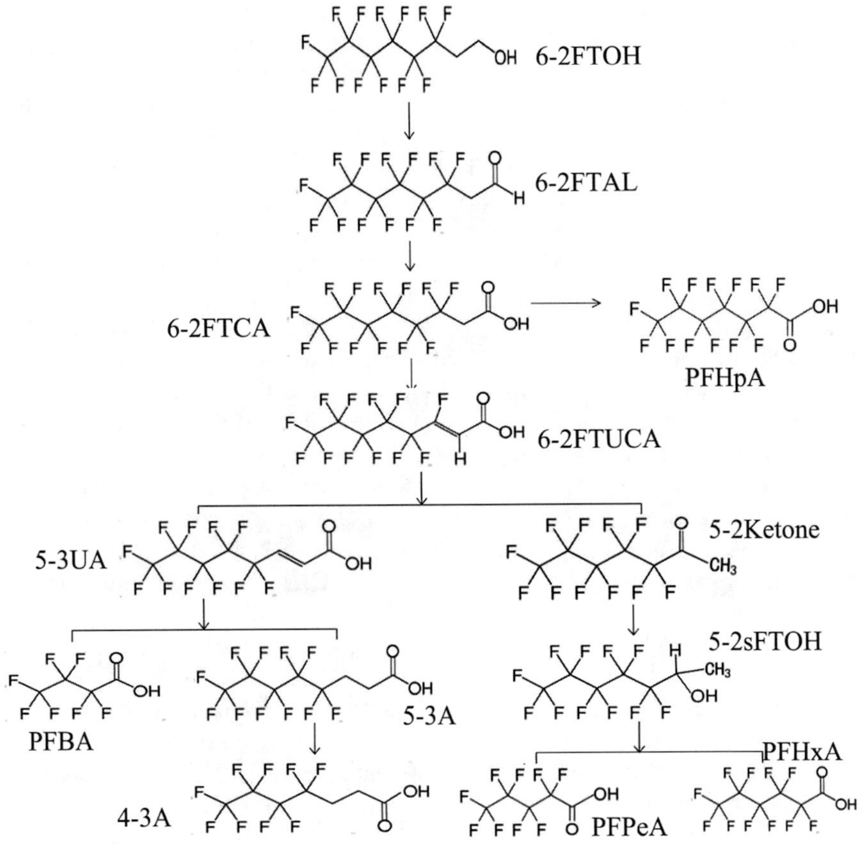

Fig. 2 Metabolic pathways of 6-2FTOH

FTOH for 30 days through dietary exposure, the half-lives for 10-2 FTCA, 10-2 FTUCA, 8-2 FTCA and 8-2 FTUCA were 3.7 ± 0.4, 2.1 ± 0.5, 3.3 and 1.3 d, respectively (Brandsma et al. 2011).

2.2 Metabolising Enzymes

There are few reports about relevant enzymes involved in the biotransformation of FTOHs. In a previous study, Martin et al. (2009) documented that ABT, a nonspecific P450 inhibitor, significantly reduced the toxicity of three FTOHs with different chain lengths at a concentration of 1 mM and no changes were shown by pyrazole and ketoconazole. Pyrazole and ketoconazole can inhibit the activity of alcohol dehydrogenase (ADH) or CYP3A, respectively. This suggests that P450 enzymes participate in the biotransformation, but CYP3A had no influence on the metabolism of FTOHs. In addition, when incubated with 8-2 FTOH, CYP2E1 plays an important role in 8-2 FTOH's biotransformation in rats. However, Li et al. (2016) reported that only CYP2C19 could metabolise 8-2 FTOH when individual incubation of recombinant human CYPs with 8-2 FTOH was conducted, and there was almost no consumption of 8-2 FTOH because omeprazole can inhibit the activity of the CYP2C19 enzyme. This difference may be related to interspecies variations.

3 Toxicity

FTOHs can cause a variety of harmful effects in animals and humans. The toxicities of FTOHs in vivo and in vitro are shown in Table 2.

Acute Toxicity In an oral acute toxicity study of 6-2 FTOH (LD_{50} = 1750 mg/kg), the changes of biological characteristics, such as appearance traits, physical behaviour, respiration, reactivity, tonic convulsions, tremors and moribund conditions were observed in rodents (Mukerji et al. 2015; Serex et al. 2014). The LC50s (±standard error) in Sprague–Dawley rats' hepatocytes for three kinds of FTOHs were 0.66 ± 0.20 (4-2 FTOH), 3.7 ± 0.54 (6-2 FTOH) and 1.4 ± 0.37 (6-2 FTOH) mM, respectively (Martin et al. 2009). According to the different chain lengths, the shortest FTOH was the most toxic (4-2 FTOH) and the longest FTOH (8-2 FTOH) was the second most toxic, while the middle FTOH (6-2 FTOH) was the least toxic.

Subchronic Toxicity An FTOH (100 and 250 mg/kg) mixture led to rats demonstrating changes in their teeth, hepatocytes, liver weights and thyroid follicular in a 90-day subchronic toxicity study (Ladics et al. 2005). In rats, 8-2 FTOH caused changes in foetal bone at a high dose of 500 mg/kg (Mylchreest et al. 2005a). For subchronic toxicity based on haematological and liver results, the not observed adverse effect level (NOAEL) of 6-2 FTOH in rats was 5 mg/kg/day (Serex et al. 2014). According to the effects on mortality, clinical observations, liver parameters, histopathology, increased body weight and white blood cells, reduction of food consumption and red blood cells, the NOAEL of 6-2 FTOH for systemic toxicity in male mice was 25 mg/kg, while in female mice it was 5 mg/kg (Mukerji et al. 2015).

Hormone Effects Rosenmai demonstrated that 8-2 FTOH resulted in a significant reduction in both androgens (testosterone and androstenedione) and progestogen (progesterone and 17-OH-progesterone) with a lowest observable effect concentration (LOEC) of 3.1 μM and caused a decrease in dehydroepiandrosterone at a concentration of 12.5 μM. An increase was found in estrone and 17 β-oestradiol levels with 8-2 FTOH exposure (12.5 μM), whereas the cortisol level was not affected (Rosenmai et al. 2013). This is contrary to previous research and the differences may be due to incubation periods (Liu et al. 2010b). 8-2 FTOH affected steroidogenesis by acting with two targets (Bzrp and CYP19 gene) causing a change in androgen and oestrogen levels (Rosenmai et al. 2013). In gene expression analysis, levels of Bzrp mRNA showed a statistically significant decrease when exposed to 12.5 and 50.0 μM 8-2 FTOH. The expression of CYP19 mRNA was

Table 2 The toxicity of FTOHs in vivo and in vitro

Species	Exposure time	Dose	Objective	Result	Reference
Rat	1, 5, 23, 90 days	6-2 FTOH (0.5 and 5.0 ppm) (1, 10 and 100 ppm) (5, 25, 125 and 250 mg/kg)	Metabolic pathways of 6-2 FTOH	In a 1-day study, the main metabolites PFBA, PFHxA, PFHpA and 5-3 Acid have a $T_{1/2}$ of 1.3–15.4 h. In 5-day, 23-day and 90-day studies, 5-3 acid has a higher concentration with $T_{1/2}$ of 20–30 days	Russell et al. (2015)
		6-2 FTOH (5, 25, 125 and 250 mg/kg/day)	Toxicity on development and reproduction	Body weight and food consumption declined. Mortality increased. Offspring survival indices were affected	O'Connor et al. (2014)
	90 days	8-2 FTOH (1, 5, 25, and 125 mg/kg)	The effects of 8-2 FTOH in the rat	Striated teeth, hepatic beta-oxidation and liver weight increased, accompanied by focal hepatic necrosis and chronic progressive nephrotoxicity	Ladics et al. (2008)
	90 days	Fluoroalkylethanol mixture (25, 100 and 250 mg/kg/day)	The subchronic toxicity of FTOHs	Body weights decreased. Liver and kidney weights, as well as hepatic β-oxidation increased. Hepatocellular hypertrophy, thyroid follicular hypertrophy, broken and absent teeth, and microscopic tooth lesions occurred	Ladics et al. (2005)
	14 days	8-2 FTOH (50, 200 and 500 mg/kg)	Developmental toxicity of 8-2 FTOH	Maternal toxicity includes maternal mortality, body weights, and clinical toxic observations. Increased foetal skeletal variations occurred	Mylchreest et al. (2005b)
	14 days	Fluoroalkylethanol mixture (50, 200 and 500 mg/kg/day)	The reproductive and developmental toxicity of FTOHs	Litter size, mortality and weights of pups decreased. Maternal body weight declined. Perineal fur staining and foetal skeletal alterations increased	Mylchreest et al. (2005a)

Mice	28 days	0.1% 8-2 FTOH(0.025, 0.05, 0.1, and 0.2%)(w/w)	Effects of 8-2 FTOH on peroxisome proliferation	The relative liver weight increased. The number of peroxisomes increased in mice hepatocyte. PFOA affected the activity of ACOX1 depending on the concentration	Kudo et al. (2005)
		6-2 FTOH (1, 5, 25, and 100 mg/kg/day)	Systemic and reproductive toxicity of mice	Liver and kidney weight gain, hepatocellular hypertrophy, Lamination of dentin and atrophy of ovarian interstitial cells occurred	Mukerji et al. (2015)
Zebrafish	4 weeks	8-2 FTOH (0, 10, 30, 90, 270 μg/L)	Endocrine and reproductive toxicity of 8-2 FTOH	Decreased the quality of eggs (eggshell, protein content and egg diameter) and hatching rates, blocked sex hormone biosynthesis, and reproduction	Liu et al. (2010a)
Rat testicular cells		PFNA (100 and 300 μM)	Potential mechanisms of testicular toxicity of PFCs	Cell viability declined after exposure to 300 μM PFNA. Cell death was increased into 17%. DNA damage increased following exposure to 300 μM PFNA	Lindeman et al. (2012)
MCF-7 Breast Cancer Cell	24 h	6-2 FTOH (0.3, 3, 30 μM) and 8-2 FTOH (0.1, 1, 10 μM)	Estrogen-like properties of FTOHs	Cells in the S-phase significantly increased. Trefoil factor 1, progesterone, oestrogen receptor, and PDZK1 upregulated. ERBB2 gene expression downregulated	Maras et al. (2006)
H295R human adrenal cortico-carcinoma cell	48 h	8-2 FTOH (1.6, 3.1, 6.3, 12.5, 25.0 and 50.0 μM)	The effects of 8-2 FTOH on sex hormone synthesis	Decreased androgens (testosterone, dehydroepiandrosterone, and androstenedione), Progesterone and 17-OH-progesterone	Rosenmai et al. (2013)
Yeast cells		6-2 FTOH and 8-2 FTOH	The estrogenic effects of FTOHs	6-2 and 8-2 FTOH increased hER-mediated transcription by interacting with the hERa or hERb ligand-binding domain	Ishibashi et al. (2007)

increased when exposed to 3.1, 12.5, and 50.0 μM of 8-2 FTOH. FTOHs show estrogenic effects by interacting with human oestrogen receptor (hER) α and β (Ishibashi et al. 2007).

Hepatotoxicity Various changes were found in mice after exposure to 6-2 FTOH at dose of 100 mg/kg, which included hypertrophy, hyperplasia and hepatocyte necrosis (Mukerji et al. 2015). Ladics also reported that 8-2 FTOH (125 mg/kg and 250 mg/kg) can increase liver weight, beta-oxidation, and hepatocyte necrosis in SD rats after 90 days of treatment (Ladics et al. 2005, 2008).

Reproductive Toxicity and Developmental Toxicity For reproductive toxicity of 6-2 FTOH, the NOAEL and LOEC were 25 and 100 mg/kg/day in mice, with decreased maternal BWG during lactation effecting oestrous cyclicity and mammary gland lesions. In rats, live births, postnatal survival, and BW gain decreased with the NOAEL (75 mg/kg) and LOEC (225 mg/kg) (Rice et al. 2020). When rats were exposed orally to different doses of 6-2 FTOH, skeletal changes in foetuses increased at 250 mg/kg/day (O'Connor et al. 2014). When zebrafish embryos were treated with 12 mg/L 6-2 FTCA, the survival rates and heart rates of zebrafish embryos decreased, while the malformation of zebrafish embryos increased, and the LC_{50} was determined to be 7.33 mg/L (Shi et al. 2017). There are a variety of metabolites of FTOHs, and most are toxic to animals and humans. The mid-metabolites may be also the one of the causes of toxicities induced by FTOHs. The toxicity of FTOH and its metabolites in species are shown in Table 3. In summary, the main metabolites of PFOA, PFNA, FTCA and FTUAL were proposed to be more toxic than FTOHs.

Thus, paying more attention to the toxicity of FTOHs is essential because of exposure through their applications in various commercial products. The consumption of different food contact products containing FTOHs may also affect long-term human health.

4 Toxicity Induced by Oxidative Stress

4.1 Generation of Oxidative Stress and ROS

Oxidative stress is caused when an imbalance between oxidation and antioxidant defences in favour of the former occurs. This means that the generation of reactive oxygen species (ROS) overwhelms their elimination. These ROS include superoxide anion radical (O^{2-}), hydroxyl radical (HO), hydroxyl anion (OH^-), hydrogen peroxide (H_2O_2), and various peroxides related to lipids, proteins and nucleic acids (Lushchak 2014). Increased levels of reactive oxygen species damage lipids, proteins, and DNA causing cell death and lipid peroxidation, protein oxidation (protein carbonyl), and oxidative DNA damage (Burton and Jauniaux 2011; Gasparovic et al. 2018; Gonzalez-Hunt et al. 2018; Mandavilli et al. 2012). Increasing evidence has suggested that oxidative stress is associated with the toxic effects of FTOHs. In vivo and in vitro PFOA-related oxidative stress studies are shown in Table 4.

Table 3 The LD50 and LC50 of FTOH and its metabolites

Assay	LD50		LC 50					
Species	Rats	Mice	Rotifer Brachionus calyciflorus	Zebrafish	Rainbow trout	Daphnia	Fathead minnow	Rats' hepatocytes
4-2 FTOH								0.66 mM[a]
6-2 FTOH	1,750 mg/kg[b]							3.7 mM[a]
8-2 FTOH								1.4 mM[a]
PFOA	189 mg/kg[c]		150.0 mg/L[d]	473 mg/L[e]		199.51[f]		
PFOS		0.579 g/kg[g]	61.8 mg/L[d]					
PFBA				13,795 mg/L[e]		>100 mg/L[h]		
8-2 FTCA					>100 mg/L[i]	2.3 mg/L[i]		
8-2 FTUA					81 mg/L[i]			
10-2 FTCA						>60 μg/L[j]		
10-2 FTUA						150 ug/L[j]		
7-3 Acid					32 mg/L[i]	2.3 mg/L[i]		
6-2 FTCA						>97.5 mg/L[i]		
6-2 FTUA						29.6 mg/L[i]		
5-3 Acid						>103 mg/L[i]		
PFPeA						>112 mg/L[i]	32 mg/L[i]	
PFHxA					>99.2 mg/L[k]	>96.5 mg/L[k]		
PFDA					32 mg/L[i]	>100 mg/L[i]		
PFNA						>100 mg/L[h]		
PFHpA						>100 mg/L[h]		

Note: The reference was cited from as following

[a]Martin et al. (2009); [b]Serex et al. (2014); [c]Olson and Andersen (1983); [d]Zhang et al. (2013a); [e]Godfrey et al. (2017); [f]Ji et al. (2008); [g]Xing et al. (2016); [h]Boudreau (2002); [i]Hoke et al. (2012); [j]Phillips et al. (2010); [k]Loveless et al. (2009)

Table 4 In vivo and in vitro PFOA-related oxidative stress studies

Species	Time of exposure	Dose	Objective	Result	Reference
Rana nigromaculata (frog)	14 days	PFOA (0, 0.01, 0.1, 0.5, or 1 mg/L)	Hepatotoxicity of PFOA and the Nrf2-ARE signalling pathway	Significantly elevated content of ROS, MDA, GLH and GSH activity. Decreased the glutathione peroxidase activities concomitantly. Increased the expression of CYP3A, Nrf2, and NQO1	Tang et al. (2018a)
Salmon	14 days	PFOA (0.2 mg/kg)	The effects of PFOA on lipid β-oxidation	Significantly increased liver MDA, CAT and SOD mRNA, and kidney CAT expression	Arukwe and Mortensen (2011)
SD rats	14 days	PFNA (1, 3, or 5 mg/kg)	The mode of PFNA-induced apoptosis	Rat spleens showed apoptosis. H_2O_2 content increased by 31.2%. SOD activity decreased by 42.2% and 55.1%	Fang et al. (2010)
Chicken embryo		PFOA (2 mg/kg) PFOA (2 mg/kg) + L-carnitine (100 mg/kg)	Developmental cardiotoxicity of PFOA and L-Carnitine mediated protection	L-carnitine protects against that the ROS and NO levels and p65 translocation was elevated by PFOA	Zhao et al. (2017a)
Mouse oocytes	24 h	PFOA(28.2 and 112.8 μM)	PFOA induce oxidative stress and lead to apoptosis	PFOA caused cell apoptosis and necrosis. ROS levels increased significantly in foetal ovaries.	Lopez-Arellano et al. (2019)
C. auratus lymphocytes	24 h	PFOA (0, 1, 10, 100 μg/L)	The cytotoxicity of PFOA and oxidative stress	The ROS and MDA content increased. SOD and GSH activity decreased. Atg 7, Atg 5 and Beclin 1 mRNA expression elevated	Tang et al. (2018b)
Cerebellar granule cells			The effects of FTOHs on ROS and cell death	8-2FTOH-induced cell death with EC_{50} of 15uM. PFOA induced ROS formation with EC_{50} of 25uM	Reistad et al. (2013)
Epithelial cells of human lungs		PFOA (0–100 μg/mL)	The cytotoxicity of PFOA to the epithelial cells of human lungs	PFOA caused the inhibition of cell viability, oxidative stress and apoptosis of A549 cells	Nong et al. (2014)
Human-hamster hybrid cell	16 days	PFOA (1–200 μM)	Mutagenicity of PFOA and the mechanisms	Concentrations of O_2^- and NO increased. ROS level increased 1.6 times. Caspase-3/7 and caspase-9 activities increased 1.31 and 1.29 times	Zhao et al. (2011)

Hepatocytes of freshwater tilapia	24 h	PFOA (0, 1, 5, 15 and 30 mg/L)	The cellular toxicology of PFOA and oxidative stress	ROS level, activities of SOD, CAT and GR were significant induced. Activities of GPx and GST, and GSH content reduced. LPO level increased	Liu et al. (2007)
HepG2 cells	3 h	PFOA (0.4 µM–2 mM)	PFOA and PFNA induce oxidative damage	PFOA increased ROS generation by 1.52-fold and PFNA increase DNA damage	Eriksen et al. (2010)
Hep G2 cells	72 h	PFOA (50–200 µmol/L)	Cytotoxic effects of PFOA when exposure to Hep G2 cells	PFOA decreased the viability of Hep G2 and induced cellular apoptosis, ROS generation. The activity of SOD, CAT, GR significantly increased while the activity of GST and GPx reduced	Hu and Hu (2009)
HepG2 cells	24 h	PFOA (50–400 uM)	Genotoxicity of PFOA on human cells	PFOA caused DNA strand breaks and induced the production of ROS and 8-OHdG	Yao and Zhong (2005)
HepG2 Cells	24 h	PFOA (200, 400 µM)	ROS and mitochondrial membrane potential in PFOA-induced apoptosis	200 µM and 400 µM of PFOA significantly increased levels of H_2O_2 and ROS	Panaretakis et al. (2001)

Peroxisome Proliferation When mice were repeatedly exposed to 8-2 FTOH (0–0.2%), liver enlargement and peroxisome proliferation (increased size and number of peroxisomes) were observed (Kudo et al. 2005). Additionally, 0.1% 8-2 FTOH can increase the number (130%) and the volume density (312%) of peroxisomes in mice. With administration of 6-2 FTOH, the consistency of hepatocellular hypertrophy and peroxisome proliferation suggested that the mechanisms underlying the hepatotoxicity may be oxidative stress (Mukerji et al. 2015; Serex et al. 2014).

Peroxisome proliferation could be caused by 8-2 FTOH through the effects of its metabolite PFOA (Kudo et al. 2005). PFOA can increase lauroyl CoA oxidase enzyme activity and lower serum cholesterol via activating peroxisome proliferator-activated receptors (PPAR), including mouse and human PPARα, and mouse PPARβ/δ (Botelho et al. 2015; Qazi et al. 2010; Takacs and Abbott 2007). PFOA and PFNA can lead to developmental toxicity and exhibit an increase in the mortality rate and slower growth through PPARα in mice (Abbott et al. 2007; Wolf et al. 2010). The PPARα signalling pathway plays an important role in PFOA-induced toxicity in rodents, but is only related to liver enzymes and the serum concentration of PFOA in humans (Li et al. 2017).

Glutathione Glutathione (GSH) is the most abundant intracellular and multifunctional antioxidant, and oxidative stress is also associated with its extensive depletion. In Sprague–Dawley rat hepatocytes treated with 1 mmol/L 8-2 FTOH for 3 h, GSH content decreased by 93% compared to a control group, and protein carbonyls and lipid peroxidation significantly increased. It was reported that 8-2 FTUAL and FTUCA can react with GSH and be further dehydrofluorinated, yielding conjugates (Martin et al. 2005). Thus, this suggests that liver toxicity is not likely due to depletion of GSH induced by 8-2 FTOH but due to its metabolites.

ROS It has been demonstrated that 8-2 FTOH-induced cell death but did not induce ROS generation, while PFOA was shown to induce an increase in ROS formation but had no effect on cell viability (Reistad et al. 2013). In vertebrate models such as TLT cells and zebrafish embryos, PFOA and PFNA both induced oxidative stress by inducing ROS production and lipid peroxidation (Rainieri et al. 2017). It was reported that ROS levels increased significantly in mouse oocytes (112.8 μM), *C. auratus* lymphocytes (1–100 μg/L), cerebellar granule cells (12 μM), human lung epithelial cells (0–100 μg/mL), HepG2 cells (0.4–200 μM) and freshwater tilapia hepatocytes (15 and 30 mg/L) when they were exposed to PFOA (Eriksen et al. 2010; Lopez-Arellano et al. 2019; Reistad et al. 2013; Tang et al. 2018b; Liu et al. 2007; Panaretakis et al. 2001; Nong et al. 2014). To explore whether PFOA would cause mutagenic effects in mammalian cells and the mechanism involved, human-hamster hybrid cells' exposure to 100 μmol/L PFOA was observed, and the levels of ROS, O_2^-, and NO increased significantly (Zhao et al. 2011). When released sperm were exposed to 400 μM PFOA in vitro, the production of ROS increased on a time-dependent basis (Lu et al. 2016). PFOA could also elevate ROS levels when fertile chicken eggs and frogs were treated with 2 mg/kg PFOA (Zhao et al. 2017a; Tang et al. 2018a).

Similarly, when zebrafish larvae were exposed to PFNA (25–100 μM), significant ROS production was observed, which could be attenuated by N-acetylcysteine. Furthermore, PFNA induced the generation of ROS by regulating the genes correlated with oxidative stress. PFNA can upregulate the expression of fatty acid binding protein 1 liver (LFABP) and uncoupling protein 2 (UCP 2) while downregulating the gene transcription of NADH dehydrogenase subunit 1 (MT-ND 1), ATP synthase F0 subunit 6 (MT-ATP 6), superoxide dismutase 1 (SOD 1), and cytochrome c oxidase subunit I (COX 1) (Liu et al. 2015a). It has been reported that UCP2, MT-ND1 and SOD 1 play an important role in reducing ROS formation by transferring an electron in the respiratory chain and destroying free superoxide radicals, respectively (Brand 2000; Valentine et al. 2005; Zelko et al. 2002).

Thus, research has indicated that ROS participates in oxidative damage and toxicity induced by FTOHs. Based on oxidative stress studies, the particular metabolites of FTOHs such as PFOA and PFNA may be responsible for the toxicity of FTOHs rather than the FTOHs themselves.

4.2 Oxidative Damage

Oxidative stress can be induced by excessive ROS production leading to damage in the balance of the antioxidant defence and oxidant system. This can damage lipids and DNA.

4.2.1 Damage to DNA

Oxidative stress tends to lead to DNA damage such as single-nucleobase lesions, DNA-protein cross-links, DNA strand breaks, tandem lesions, and nucleobase intrastrand cross-links (Dizdaroglu and Jaruga 2012; Gonzalez-Hunt et al. 2018; Yu et al. 2016). There are various methods of measuring DNA damage, and the comet assay is one of the most widely used techniques. The comet assay often measures the level of 8-hydroxyguanine (8-oxoguanine or 8-OH-dG) which is the main product of DNA oxidation, and its concentration is a measurement of oxidative stress (Cooke et al. 2003).

When human hepatoblastoma HepG2 cells were treated with PFOA (150 μM), exposure to PFOA induced apoptosis and DNA breakage as well as perturbations in the cell cycle (Shabalina et al. 1999). A tenfold increase in 8-OH-dG levels was observed when HepG2 cells were exposed to 400 μM PFOA for 3 h (Yao and Zhong 2005). Through immunocytochemical detection, it was observed that PFOA could also increase the 8-OHdG levels in human epidermal HaCaT cells after exposure to 50 μM PFOA (Peropadre et al. 2018). After male rats were treated with a single 100 mg/kg dose of PFOA for 3, 5, and 8 days, liver weight and the generation of 8-OH-dG in liver DNA were found to have significantly increased (Takagi et al. 1991).

These studies demonstrated that the metabolites of FTOHs can induce DNA damage by oxidative stress, indicating that the toxicity of FTOHs may be caused by oxidative stress that is induced by its metabolites.

4.2.2 Damage to Lipids

Mass production of lipid peroxidation products is caused by the accumulation of reactive species induced by oxidative stress. Lipid peroxidation changes lipid–lipid interactions, lipid diffusion, electron transport and even causes cell death (Ali et al. 2018; Catala and Diaz 2016). MDA, one of the major lipid peroxide degradation products, is a widely used marker of lipid peroxidation (Niki 2014; Roberts 1998). MDA is measured via the quantification (by absorbance) of a chromophore that forms from MDA and thiobarbituric acid (Lykkesfeldt 2007).

After cultured freshwater tilapia hepatocytes were treated with PFOA, the MDA content increased. Depending on the dose, it increased by 53.8%, 59.6%, and 90.5% at concentrations of 5, 15, and 30 mg/L, respectively (Liu et al. 2007). In zebrafish larvae treated with 15 mg/L PFNA, MDA content increased by 100.9% (Yang et al. 2014b). Arukwe and Mortensen (2011) reported that when salmon fish were exposed to PFOA and PFOS at a dose of 0.2 mg/kg, the liver MDA content significantly increased by two- and threefold.

To explore whether PFOA exposure led to oxidative stress in mouse liver, the content of MDA and H_2O_2 significantly increased after mice were treated with 5 and 10 mg/kg PFOA for 14 days ($P < 0.05$) (Yang et al. 2014a). After BABL/c male mice were treated with 1.25 and 5 mg/kg PFOA in vivo, the MDA content increased compared with a control group (Lu et al. 2016).

In vitro and in vivo studies demonstrated that oxidative stress often results in lipid peroxidation. PFOA could do oxidative damage to lipids in the kidney and liver, suggesting that liver and kidney toxicity induced by FTOHs might be due to oxidative stress caused by their metabolites.

4.3 Alterations in Antioxidant Status

A complex antioxidant defence grid acts against free radicals to protect cells from damage. The defence antioxidants include superoxide dismutase (SOD), catalase (CAT), glutathione peroxidase (GP_X), glutathione reductase (GR), GSH and glutathione S-transferase (GST). These are important for inhibiting the generation of free radicals and can also serve as redox biomarkers (Farhat et al. 2018; Ighodaro and Akinloye 2018).

After salmon were fed PFOA or PFOS (0.2 mg/kg), both liver SOD and CAT mRNA significantly increased in the first, and they decreased after 8 days (Arukwe and Mortensen 2011). SOD activity significantly increased when zebrafish larvae were treated with 1 and 5 mg/L PFNA, but a decrease in SOD and CAT activity occurred at a dose of 15 mg/L (Yang et al. 2014b).

When freshwater tilapia hepatocytes were treated with PFOA for 24 h, SOD, CAT, and GR activities increased; while GPx, GST activities, and GSH content decreased (Liu et al. 2007). In PFOA-treated Hep G2 cells, SOD, CAT, and GR activities increased by 53.9, 55.9, and 91.0% at a dose of 200 μmol/L, respectively. GPx, GST activities, and GSH content decreased by 45.9%, 39.9% and 47.9% when treated with 200 μmol/L PFOA (Hu and Hu 2009). Exposure to PFOA (15 mg/L) for 2 weeks resulted in reduction in GR activity, the GSH and ATP levels, and an increase in GPx, GST, ATPase activity, and the glutathione disulphide (GSSG) contents (Zhang et al. 2019). This is in contrast to prior reports, which demonstrated that PFOA induced oxidative stress by decreasing GSH content and GPx and GST activities (Liu et al. 2007; Liu et al. 2008). The increase in GPx content and GST activity may have occurred when the antioxidant systems were disrupted by damage from excessive ROS and MDA production induced by PFOA.

The metabolites of FTOHs can significantly alter the status of antioxidant enzymes. This suggests that the toxic effects of FTOHs in vivo and in vitro involve the antioxidant status imbalance caused by its metabolites. However, the toxic effects of FTOHs related to the imbalance of antioxidant status and their impact on factors mediating the antioxidant enzymes remain unclear and need to be further investigated in the future.

4.4 Cell Signalling

To investigate the toxic mechanisms mediated by PFOA in human liver, Tian et al. studied the possible effects of DNA methylation induced by PFOA. They used L02 cells and found that PFOA increased the methylation of three genes (GSTP, PRB, and ERα) involved in promoting methylation, although there are two sites (SP1) to protect from methylation in the GSTP region. PFOA also significantly increased mRNA transcript levels of DNA methyltransferases (DNMT3A) that catalyse DNA methylation. But PFOA had no obvious effect on global genomic methylation. This suggests that PFOA inhibits SP1 binding to the GSTP promoter that increases DNA methylation, leading to the toxic response (Tian et al. 2012). PFOA can increase the p53 mRNA expression levels in L-02 cells, indicating that PFOA induces apoptosis through the p53 signalling pathway. This suggests that p53 was activated by increasing ROS production and inhibiting the expression of GRP78, HSP27, CTSD, and hnRNPC. The expression of Bax was then upregulated, the expression of Bcl-2 was downregulated, and mitochondrial cytochrome c was released to the cytosol, which can lead to cell apoptosis by activating the caspases (caspase-9 and -3) (Huang et al. 2013).

Box, Bcl-2 and Caspase-3 (involved in the regulation of apoptosis) and apoptosis-inducing factor (AIF) can initiate caspase-independent apoptosis. Bcl-2 is an inhibitor of Bax. It was reported that Bcl-2 expression was downregulated in zebrafish larvae exposed to PFNA, and a 1.8- and 2.1-fold increase in AIF was observed at doses of 0.5 and 5 mg/L (Yang et al. 2014b). The levels of SOD and Bcl-2 were significantly reduced in rat spleens, and AIF significantly increased with exposure to PFNA at concentrations of 3 and 5 mg/kg/day. However, Bax levels exhibited no significant difference (Fang et al. 2010). Lee et al. (2014) confirmed that PFHxS can induce apoptosis in neuronal cells through NMDA receptor-mediated ERK activation.

Atg 5 and Atg 7 play important roles in the initiation of autophagy to form the pre-autophagosomal structure. LC3 was associated with autophagy vesicles, and then LC3-I was converted to LC3-II (Nitta et al. 2019). When upregulating the expression of Beclin 1, autophagy was stimulated (Liang et al. 1999). After *C. auratus* lymphocytes were incubated with PFOA, the upregulation of Atg 5, Atg 7, and Beclin 1 was observed, an increase of ROS production and MDA content also occurred, and SOD and GSH activity decreased. This suggests that PFOA leads to autophagy by inducing oxidative stress (Tang et al. 2018b).

In frogs, a significant increase in CYP3A, Nrf2, HO-1 and NQO1 expression occurred after frogs were exposed to PFOA. Hepatotoxicity may be enhanced by upregulating the expression of CYP3A but inhibited by upregulating the expression of NQO1, HO-1, and Nrf2. The results showed the role of the Nrf2-ARE pathway in regulating hepatotoxicity induced by PFOA (Tang et al. 2018a).

The signalling pathways including PPAR, apoptosis, p53, and autophagy were involved in the toxicity induced by FTOHs and its metabolites. Notably, these pathways may be closely related to ROS production, indicating that oxidative stress may play an important role in the toxic effects induced by FTOHs.

5 Conclusion and Prospects

The widespread and rapidly increasing usage of FTOHs in recent years has led to increasing concern about their harmful impact on animal and human health. Toxicological research in this area is therefore of the utmost importance to assess the risks of FTOHs in animals and humans. This review contributes to an improved understanding of the effects of FTOH-induced toxicity while also illuminating the role of oxidative stress and the toxicity of FTOH metabolites. It is important to guard against FTOH-induced toxicity and protect animal and human health.

Although the toxicity of FTOHs is low, they can cause adverse effects through their metabolite products. For occupational exposure and accumulation, studying the toxicity of the metabolites and the mechanism is important. FTOHs can degrade into many different metabolites with different chain lengths. The combined toxicity of FTOHs and their toxic metabolites may increase the toxic effects of FTOHs. In humans, PFOA can decrease efficiency of immunisations, but increase the prevalence of fever, cold, and gastroenteritis in children ages 1–4 years when the median serum level of PFOA was 1.68 ng/mL (Dalsager et al. 2016). PFNA and PFDA are reportedly linked to hearing impairment when serum concentrations of PFNA and PFDA were 1.8 ng/mL and 0.5 ng/mL (Ding and Park 2020).

More attention needs to be paid to the relationship between metabolites and protoplast toxicity as well as to methods of reducing or offsetting their toxicity, while minimising the negative impact on the environment. Thus, further studies are required regarding the effect of enzymes and FTOHs reacting with biomolecules in the body to ascertain the mechanisms of interactions and modes of action. FTOHs and their metabolites usually do not exist alone and tend to interact with other chemicals in organisms. These interactions may affect their function in the body and enhance the toxicities of these chemicals. The multiple chemical interactions in the environment and the body are worthy of further study.

Oxidative stress induced by FTOHs and other metabolites has been poorly investigated. Therefore, it is necessary to conduct further studies of oxidative stress that occurs in the toxic mechanism of other metabolites and FTOHs. We hypothesise (Fig. 3) that after 8-2 FTOH degrades into the final metabolite PFOA, PFOA leads to

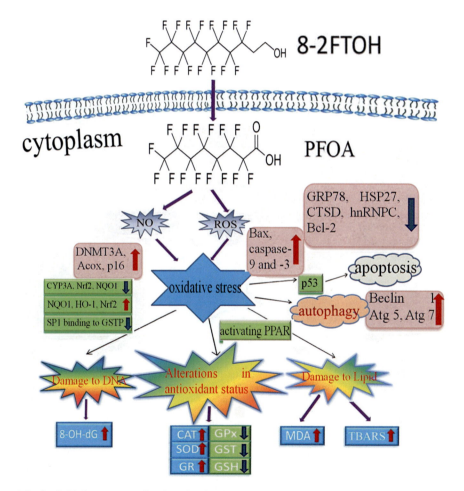

Fig. 3 Oxidative stress-mediated mode of action proposed for 8-2FTOH the preventive effect of different compounds

oxidative stress by producing excessive ROS and NO, in which the PPAR, apoptosis, and autophagy signalling pathway are involved. The PPARα signalling pathway may be an important mechanism of PFOA-induced toxic effects in rodents but may not correlate with human health hazards except increasing liver enzymes. The mitochondrial signalling pathway warrants further research (Li et al. 2017). Future studies could focus on apoptosis and the autophagy pathway that is more likely to impact human health.

Acknowledgments This work was supported by Natural Science Foundation of China (NSFC, 31572570) and Risk Assessment of Unknown and known hazard factors of livestock and poultry products (GJFP2019007).

Conflict of Interest The authors declare that they have no conflict of interest.

References

Abbott BD, Wolf CJ, Schmid JE, Das KP, Zehr RD, Helfant L, Nakayama S, Lindstrom AB, Strynar MJ, Lau C (2007) Perfluorooctanoic acid-induced developmental toxicity in the mouse is dependent on expression of peroxisome proliferator-activated receptor-alpha. Toxicol Sci 98:571–581. https://doi.org/10.1093/toxsci/kfm110

Ali H, Assiri MA, Shearn CT, Fritz KS (2018) Lipid peroxidation derived reactive aldehydes in alcoholic liver disease. Curr Opinion Toxicol. https://doi.org/10.1016/j.cotox.2018.10.003

Arukwe A, Mortensen AS (2011) Lipid peroxidation and oxidative stress responses of salmon fed a diet containing perfluorooctane sulfonic- or perfluorooctane carboxylic acids. Comp Biochem Phys C 154:288–295. https://doi.org/10.1016/j.cbpc.2011.06.012

Botelho SC, Saghafian M, Pavlova S, Hassan M, DePierre JW, Abedi-Valugerdi M (2015) Complement activation is involved in the hepatic injury caused by high-dose exposure of mice to perfluorooctanoic acid. Chemosphere 129:225–231. https://doi.org/10.1016/j.chemosphere.2014.06.093

Boudreau TM (2002) Toxicity of perfluorinated organic anions to selected organisms under laboratory and field conditions. Master's Thesis. Department of Environmental Biology. University of Guelph, Guelph, p 137

Brand MD (2000) Uncoupling to survive? The role of mitochondrial inefficiency in ageing. Exp Gerontol 35:811–820. https://doi.org/10.1016/S0531-5565(00)00135-2

Brandsma SH, Smithwick M, Solomon K, Small J, de Boer J, Muir DC (2011) Dietary exposure of rainbow trout to 8:2 and 10:2 fluorotelomer alcohols and perfluorooctanesulfonamide: Uptake, transformation and elimination. Chemosphere 82:253–258. https://doi.org/10.1016/j.chemosphere.2010.09.050

Buck RC, Franklin J, Berger U, Conder JM, Cousins IT, de Voogt P, Jensen AA, Kannan K, Mabury SA, van Leeuwen SPJ (2011) Perfluoroalkyl and polyfluoroalkyl substances in the environment: terminology, classification, and origins. Integr Environ Asses 7:513–541. https://doi.org/10.1002/ieam.258

Burton GJ, Jauniaux E (2011) Oxidative stress. Best Pract Res Clin Obstet Gynaecol 25:287–299. https://doi.org/10.1016/j.bpobgyn.2010.10.016

Butt CM, Berger U, Bossi R, Tomy GT (2010a) Levels and trends of poly- and perfluorinated compounds in the arctic environment. Sci Total Environ 408:2936–2965. https://doi.org/10.1016/j.scitotenv.2010.03.015

Butt CM, Muir DCG, Mabury SA (2010b) Elucidating the pathways of polyand perfluorinated acid formation in rainbow trout. Environ Sci Technol 44:4973–4980

Butt CM, Muir DCG, Mabury SA (2014) Biotransformation pathways of fluorotelomer-based polyfluoroalkyl substances: a review. Environ Toxicol Chem 33:243–267. https://doi.org/10.1002/etc.2407

Catala A, Diaz M (2016) Editorial: impact of lipid peroxidation on the physiology and pathophysiology of cell membranes. Front Physiol 7. https://doi.org/10.3389/fphys.2016.00423

Cooke MS, Evans MD, Dizdaroglu M, Lunec J (2003) Oxidative DNA damage: mechanisms, mutation, and disease. FASEB J 17:1195–1214. https://doi.org/10.1096/fj.02-0752rev

Cui Y, Gao J, Tao Y, Wang Y, Peng D, Yuan Z, Chen D (2016) Research progress on biotransformation of perfluoroalkyl and polyfluoroalkyl substances in animals. Huanjing Huaxue Environ Chem 35:1994–2007. https://doi.org/10.7524/j.issn.0254-6108.2016.10.2016031501

Dagnino S, Strynar MJ, McMahen RL, Lau CS, Ball C, Garantziotis S (2016) Identification of biomarkers of exposure to FTOHs and PAPs in humans using a targeted and nontargeted analysis approach. Environ Sci Technol 50(18):10216–10225. https://doi.org/10.1021/acs.est.6b01170

Dalsager L, Christensen N, Husby S, Kyhl H, Nielsen F, Høst A (2016) Association between prenatal exposure to perfluorinated compounds and symptoms of infections at age 1–4 years among 359 children in the Odense Child Cohort. Environ Int 96:58–64. https://doi.org/10.1016/j.envint.2016.08.026

Ding N, Park SK (2020) Perfluoroalkyl substances exposure and hearing impairment in US adults. Environ Res 187:109686. https://doi.org/10.1016/j.envres.2020.109686

Dinglasan MJA, Ye Y, Edwards EA, Mabury SA (2004) Fluorotelomer alcohol biodegradation yields poly- and perfluorinated acids. Environ Sci Technol 38:2857–2864. https://doi.org/10.1021/es0350177

Dizdaroglu M, Jaruga P (2012) Mechanisms of free radical-induced damage to DNA. Free Radic Res 46:382–419. https://doi.org/10.3109/10715762.2011.653969

Ellis DA, Martin JW, De Silva AO, Mabury SA, Hurley MD, Andersen MPS, Wallington TJ (2004) Degradation of fluorotelomer alcohols: a likely atmospheric source of perfluorinated carboxylic acids. Environ Sci Technol 38:3316–3321. https://doi.org/10.1021/es049860w

Eriksen KT, Raaschou-Nielsen O, Sørensen M, Roursgaard M, Loft S, Møller P (2010) Genotoxic potential of the perfluorinated chemicals PFOA, PFOS, PFBS, PFNA and PFHxA in human HepG2 cells. Mutat Res-Gen Toxicol Environ 700:39–43. https://doi.org/10.1016/j.mrgentox.2010.04.024

Fang X, Feng Y, Wang J, Dai J (2010) Perfluorononanoic acid-induced apoptosis in rat spleen involves oxidative stress and the activation of caspase-independent death pathway. Toxicology 267:54–59. https://doi.org/10.1016/j.tox.2009.10.020

Farhat Z, Browne RW, Bonner MR, Tian L, Deng F, Swanson M, Mu L (2018) How do glutathione antioxidant enzymes and total antioxidant status respond to air pollution exposure? Environ Int 112:287–293. https://doi.org/10.1016/j.envint.2017.12.033

Fasano WJ, Carpenter SC, Gannon SA, Snow TA, Stadler JC, Kennedy GL, Buck RC, Korzeniowski SH, Hinderliter PM, Kemper RA (2006) Absorption, distribution, metabolism, and elimination of 8-2 fluorotelomer alcohol in the rat. Toxicol Sci 91:341–355. https://doi.org/10.1093/toxsci/kfj160

Fasano WJ, Sweeney LM, Mawn MP, Nabb DL, Szostek B, Buck RC, Gargas ML (2009) Kinetics of 8-2 fluorotelomer alcohol and its metabolites, and liver glutathione status following daily oral dosing for 45 days in male and female rats. Chem Biol Interact 180:281–295. https://doi.org/10.1016/j.cbi.2009.03.015

Fraser AJ, Webster TF, Watkins DJ, Nelson JW, Stapleton HM, Calafat AM (2012) Polyfluorinated compounds in serum linked to indoor air in office environments. Environ Sci Technol 46 (2):1209–1215

Fraser AJ, Webster TF, Watkins DJ, Strynar MJ, Kato K, Calafat AM, Vieira VM, McClean MD (2013) Polyfluorinated compounds in dust from homes, offices, and vehicles as predictors of concentrations in office workers' serum. Environ Int 60:128–136. https://doi.org/10.1016/j.envint.2013.08.012

Froemel T, Knepper TP (2010) Biodegradation of fluorinated alkyl substances. In: Whitacre DM, DeVoogt P (eds) Reviews of environmental contamination and toxicology, vol 208: perfluorinated alkylated substances. Springer Science & Business Media, Berlin, pp 161–177

Gannon SA, Nabb DL, Snow TA, Mawn MP, Serex TL, Buck RC, Loveless SE (2012) P34—In vitro metabolism of 6-2 fluorotelomer alcohol in rat, mouse, and human hepatocytes. Reprod Toxicol 33:610–611. https://doi.org/10.1016/j.reprotox.2011.11.068

Gasparovic AC, Zarkovic N, Bottari SP (2018) Biomarkers of nitro-oxidation and oxidative stress. Curr Opin Toxicol 7:73–80. https://doi.org/10.1016/j.cotox.2017.10.002

Godfrey A, Abdel-moneim A, Sepúlveda MS (2017) Acute mixture toxicity of halogenated chemicals and their next generation counterparts on zebrafish embryos. Chemosphere 181:710–712. https://doi.org/10.1016/j.chemosphere.2017.04.146

Gonzalez-Hunt CP, Wadhwa M, Sanders LH (2018) DNA damage by oxidative stress: measurement strategies for two genomes. Curr Opin Toxicol 7:87–94. https://doi.org/10.1016/j.cotox.2017.11.001

Hekster FM, Laane R, de Voogt P (2003) Environmental and toxicity effects of perfluoroalkylated substances. Rev Environ Contam Toxicol 179(179):99–121. https://doi.org/10.1007/0-387-21731-2_4

Henderson WM, Smith MA (2007) Perfluorooctanoic acid and perfluorononanoic acid in fetal and neonatal mice following in utero exposure to 8-2 fluorotelomer alcohol. Toxicol Sci 95:452–461. https://doi.org/10.1093/toxsci/kfl162

Heydebreck F, Tang J, Xie Z, Ebinghaus R (2016) Emissions of per- and polyfluoroalkyl substances in a textile manufacturing plant in China and their relevance for workers' exposure. Environ Sci Technol 50:10386–10396. https://doi.org/10.1021/acs.est.6b03213

Hoke RA, Bouchelle LD, Ferrell BD, Buck RC (2012) Comparative acute freshwater hazard assessment and preliminary PNEC development for eight fluorinated acids. Chemosphere 87:725–733. https://doi.org/10.1016/j.chemosphere.2011.12.066

Hu X-Z, Hu D-C (2009) Effects of perfluorooctanoate and perfluorooctane sulfonate exposure on hepatoma Hep G2 cells. Arch Toxicol 83:851–861. https://doi.org/10.1007/s00204-009-0441-z

Huang Q, Zhang J, Martin FL, Peng S, Tian M, Mu X, Shen H (2013) Perfluorooctanoic acid induces apoptosis through the p53-dependent mitochondrial pathway in human hepatic cells: a proteomic study. Toxicol Lett 223:211–220. https://doi.org/10.1016/j.toxlet.2013.09.002

Huang MC, Robinson VG, Waidyanatha S, Dzierlenga AL, DeVito MJ, Eifrid MA (2019) Toxicokinetics of 8:2 fluorotelomer alcohol (8:2-FTOH) in male and female Hsd:Sprague Dawley SD rats after intravenous and gavage administration. Toxicol Rep 6:924–932. https://doi.org/10.1016/j.toxrep.2019.08.009

Ighodaro OM, Akinloye OA (2018) First line defence antioxidants-superoxide dismutase (SOD), catalase (CAT) and glutathione peroxidase (GPX): their fundamental role in the entire antioxidant defence grid. Alex J Med 54:287–293. https://doi.org/10.1016/j.ajme.2017.09.001

Ishibashi H, Ishida H, Matsuoka M, Tominaga N, Arizono K (2007) Estrogenic effects of fluorotelomer alcohols for human estrogen receptor isoforms alpha and beta in vitro. Biol Pharm Bull 30:1358–1359. https://doi.org/10.1248/bpb.30.1358

Ji K, Kim Y, Oh S, Ahn B, Jo H, Choi K (2008) Toxicity of perfluorooctane sulfonic acid and perfluorooctanoic acid on freshwater macroinvertebrates (Daphnia magna and *Moina macrocopa*) and fish (*Oryzias latipes*). Environ Toxicol Chem 27:2159–2168. https://doi.org/10.1897/07-523.1

Kotthoff M, Muller J, Jurling H, Schlummer M, Fiedler D (2015) Perfluoroalkyl and polyfluoroalkyl substances in consumer products. Environ Sci Pollut R 22:14546–14559. https://doi.org/10.1007/s11356-015-4202-7

Kudo N, Iwase Y, Okayachi H, Yamakawa Y, Kawashima Y (2005) Induction of hepatic peroxisome proliferation by 8-2 telomer alcohol feeding in mice: formation of perfluorooctanoic acid in the liver. Toxicol Sci 86:231–238. https://doi.org/10.1093/toxsci/kfl191

Ladics GS, Stadler JC, Makovec GT, Everds NE, Buck RC (2005) Subchronic toxicity of a fluoroalkylethanol mixture in rats. Drug Chem Toxicol 28:135–158. https://doi.org/10.1081/dct-200052506

Ladics GS, Kennedy GL, O'Connor J, Everds N, Malley LA, Frame SR, Gannon S, Jung R, Roth T, Iwai H, Shin-Ya S (2008) 90-day oral gavage toxicity study of 8-2 fluorotelomer alcohol in rats. Drug Chem Toxicol 31:189–216. https://doi.org/10.1080/01480540701873103

Lai SC, Song JW, Song TL, Huang ZJ, Zhang YY, Zhao Y, Liu GC, Zheng JY, Mi WY, Tang JH, Zou SC, Ebinghaus R, Xie ZY (2016) Neutral polyfluoroalkyl substances in the atmosphere over the northern South China Sea. Environ Pollut 214:449–455. https://doi.org/10.1016/j.envpol.2016.04.047

Lee YJ, Choi S-Y, Yang JH (2014) NMDA receptor-mediated ERK 1/2 pathway is involved in PFHxS-induced apoptosis of PC12 cells. Sci Total Environ 491-492:227–234. https://doi.org/10.1016/j.scitotenv.2014.01.114

Li Z-M, Guo L-H, Ren X-M (2016) Biotransformation of 8:2 fluorotelomer alcohol by recombinant human cytochrome P450s, human liver microsomes and human liver cytosol. Environ Sci Proc Impacts 18:538–546. https://doi.org/10.1039/c6em00071a

Li K, Gao P, Xiang P, Zhang X, Cui X, Ma LQ (2017) Molecular mechanisms of PFOA-induced toxicity in animals and humans: implications for health risks. Environ Int 99:43–54. https://doi.org/10.1016/j.envint.2016.11.014

Liang XH, Jackson S, Seaman M, Brown K, Kempkes B, Hibshoosh H, Levine B (1999) Induction of autophagy and inhibition of tumorigenesis by beclin 1. Nature 402:672. https://doi.org/10.1038/45257

Lindeman B, Maass C, Duale N, Gützkow KB, Brunborg G, Andreassen Å (2012) Effects of per- and polyfluorinated compounds on adult rat testicular cells following in vitro exposure. Reprod Toxicol 33:531–537. https://doi.org/10.1016/j.reprotox.2011.04.001

Liu C, Yu K, Shi X, Wang J, Lam PKS, Wu RSS, Zhou B (2007) Induction of oxidative stress and apoptosis by PFOS and PFOA in primary cultured hepatocytes of freshwater tilapia (*Oreochromis niloticus*). Aquat Toxicol 82:135–143. https://doi.org/10.1016/j.aquatox.2007.02.006

Liu Y, Wang J, Wei Y, Zhang H, Xu M, Dai J (2008) Induction of time-dependent oxidative stress and related transcriptional effects of perfluorododecanoic acid in zebrafish liver. Aquat Toxicol 89:242–250. https://doi.org/10.1016/j.aquatox.2008.07.009

Liu C, Deng J, Yu L, Ramesh M, Zhou B (2010a) Endocrine disruption and reproductive impairment in zebrafish by exposure to 8:2 fluorotelomer alcohol. Aquat Toxicol 96:70–76. https://doi.org/10.1016/j.aquatox.2009.09.012

Liu C, Zhang X, Chang H, Jones P, Wiseman S, Naile J, Hecker M, Giesy JP, Zhou B (2010b) Effects of fluorotelomer alcohol 8:2 FTOH on steroidogenesis in H295R cells: targeting the cAMP signalling cascade. Toxicol Appl Pharm 247:222–228. https://doi.org/10.1016/j.taap.2010.06.016

Liu H, Sheng N, Zhang W, Dai J (2015a) Toxic effects of perfluorononanoic acid on the development of Zebrafish (*Danio rerio*) embryos. J Environ Sci 32:26–34. https://doi.org/10.1016/j.jes.2014.11.008

Liu X, Guo Z, Folk EE IV, Roache NF (2015b) Determination of fluorotelomer alcohols in selected consumer products and preliminary investigation of their fate in the indoor environment. Chemosphere 129:81–86

Lopez-Arellano P, Lopez-Arellano K, Luna J, Flores D, Jimenez-Salazar J, Gavia G, Tetetitla M, Rodriguez JJ, Dominguez A, Casas E, Bahena I, Betancourt M, Gonzalez C, Ducolomb Y, Bonilla E (2019) Perfluorooctanoic acid disrupts gap junction intercellular communication and induces reactive oxygen species formation and apoptosis in mouse ovaries. Environ Toxicol 34:92–98. https://doi.org/10.1002/tox.22661

Loveless SE, Slezak B, Serex T, Lewis J, Mukerji P, O'Connor JC, Donner EM, Frame SR, Korzeniowski SH, Buck RC (2009) Toxicological evaluation of sodium perfluorohexanoate. Toxicology 264:32–44. https://doi.org/10.1016/j.tox.2009.07.011

Lu Y, Pan Y, Sheng N, Zhao AZ, Dai J (2016) Perfluorooctanoic acid exposure alters polyunsaturated fatty acid composition, induces oxidative stress and activates the AKT/AMPK pathway in mouse epididymis. Chemosphere 158:143–153. https://doi.org/10.1016/j.chemosphere.2016. 05.071

Lushchak VI (2014) Free radicals, reactive oxygen species, oxidative stress and its classification. Chem Biol Interact 224:164–175. https://doi.org/10.1016/j.cbi.2014.10.016

Lykkesfeldt J (2007) Malondialdehyde as biomarker of oxidative damage to lipids caused by smoking. Clin Chim Acta 380:50–58. https://doi.org/10.1016/j.cca.2007.01.028

Mandavilli BS, Aggeler RJ, Singh U, Kang HC, Gee K, Agnew B, Janes MS (2012) Cell-based analysis of oxidative stress, lipid peroxidation and lipid peroxidation-derived protein modifications using fluorescence microscopy. Free Radical Bio Med 53:S136–S136. https://doi.org/10. 1016/j.freeradbiomed.2012.10.371

Maras M, Vanparys C, Muylle F, Robbens J, Berger U, Barber JL, Blust R, De Coen W (2006) Estrogen-like properties of fluorotelomer alcohols as revealed by MCF-7 breast cancer cell proliferation. Environ Health Perspect 114:100–105. https://doi.org/10.1289/ehp.8149

Martin JW, Mabury SA, O'Brien PJ (2005) Metabolic products and pathways of fluorotelomer alcohols in isolated rat hepatocytes. Chem Biol Interact 155:165–180. https://doi.org/10.1016/j. cbi.2005.06.007

Martin JW, Chan K, Mabury SA, O'Brien PJ (2009) Bioactivation of fluorotelomer alcohols in isolated rat hepatocytes. Chem Biol Interact 177:196–203. https://doi.org/10.1016/j.cbi.2008. 11.001

Mukerji P, Rae JC, Buck RC, O'Connor JC (2015) Oral repeated-dose systemic and reproductive toxicity of 6:2 fluorotelomer alcohol in mice. Toxicol Rep 2:130–143. https://doi.org/10.1016/j. toxrep.2014.12.002

Mylchreest E, Ladics G, Munley S, Buck R, Stadler J (2005a) Evaluation of the reproductive and developmental toxicity of a fluoroalkylethanol mixture. Drug Chem Toxicol 28:159–175. https://doi.org/10.1081/dct-52518

Mylchreest E, Munley SM, Kennedy GL Jr (2005b) Evaluation of the developmental toxicity of 8-2 telomer B alcohol. Drug Chem Toxicol 28:315–328. https://doi.org/10.1081/DCT-200064491

Nabb DL, Szostek B, Himmelstein MW, Mawn MP, Gargas ML, Sweeney LM, Stadler JC, Buck RC, Fasano WJ (2007) In vitro metabolism of 8-2 fluorotelomer alcohol: interspecies comparisons and metabolic pathway refinement. Toxicol Sci 100:333–344. https://doi.org/10.1093/ toxsci/kfm230

Niki E (2014) Biomarkers of lipid peroxidation in clinical material. BBA-Gen Subjects 1840:809–817. https://doi.org/10.1016/j.bbagen.2013.03.020

Nilsson H, Karrman A, Rotander A, van Bavel B, Lindstrom G, Westberg H (2010a) Inhalation exposure to fluorotelomer alcohols yield perfluorocarboxylates in human blood? Environ Sci Technol 44:7717–7722. https://doi.org/10.1021/es101951t

Nilsson H, Karrman A, Westberg H, Rotander A, Van Bavel B, Lindstrom G (2010b) A time trend study of significantly elevated perfluorocarboxylate levels in humans after using fluorinated ski wax. Environ Sci Technol 44:2150–2155. https://doi.org/10.1021/es9034733

Nilsson H, Karrman A, Rotander A, van Bavel B, Lindstrom G, Westberg H (2013) Biotransformation of fluorotelomer compound to perfluorocarboxylates in humans. Environ Int 51:8–12. https://doi.org/10.1016/j.envint.2012.09.001

Nitta A, Hori K, Tanida I, Igarashi A, Deyama Y, Ueno T, Kominami E, Sugai M, Aoki K (2019) Blocking LC3 lipidation and ATG12 conjugation reactions by ATG7 mutant protein containing C572S. Biochem Biophys Res Commun 508:521–526. https://doi.org/10.1016/j.bbrc.2018.11. 158

Nong R, Yang C, Li Y, Guo C, Dang Z (2014) Cytotoxicity of PFOA to the epithelial cells of human lungs. J Saf Environ 14:333–337

O'Connor JC, Munley SM, Serex TL, Buck RC (2014) Evaluation of the reproductive and developmental toxicity of 6:2 fluorotelomer alcohol in rats. Toxicology 317:6–16. https://doi. org/10.1016/j.tox.2014.01.002

Olson CT, Andersen ME (1983) The acute toxicity of perfluorooctanoic and perfluorodecanoic acids in male rats and effects on tissue fatty acids. Toxicol Appl Pharm 70(3):362–372. https:// doi.org/10.1016/0041-008X(83)90154-0

Panaretakis T, Shabalina IG, Grandér D, Shoshan MC, DePierre JW (2001) Reactive oxygen species and mitochondria mediate the induction of apoptosis in human hepatoma HepG2 cells by the rodent peroxisome proliferator and hepatocarcinogen, perfluorooctanoic acid. Toxicol Appl Pharm 173:56–64. https://doi.org/10.1006/taap.2001.9159

Perkins RG, Butenhoff JL, Kennedy GL, Palazzolo MJ (2004) 13-week dietary toxicity study of ammonium perfluorooctanoate (APFO) in male rats. Drug Chem Toxicol 27:361–378. https://doi.org/10.1081/dct-200039773

Peropadre A, Fernández Freire P, Hazen MJ (2018) A moderate exposure to perfluorooctanoic acid causes persistent DNA damage and senescence in human epidermal HaCaT keratinocytes. Food Chem Toxicol 121:351–359. https://doi.org/10.1016/j.fct.2018.09.020

Phillips MM, Dinglasan-Panlilio MJ, Mabury SA, Solomon KR, Sibley PK (2010) Chronic toxicity of fluorotelomer acids to Daphnia magna and Chironomus dilutus. Environ Toxicol Chem 29:1123–1131. https://doi.org/10.1002/etc.141

Qazi MR, Abedi MR, Nelson BD, DePierre JW, Abedi-Valugerdi M (2010) Dietary exposure to perfluorooctanoate or perfluorooctane sulfonate induces hypertrophy in centrilobular hepatocytes and alters the hepatic immune status in mice. Int Immunopharmacol 10:1420–1427. https://doi.org/10.1016/j.intimp.2010.08.009

Rainieri S, Conlledo N, Langerholc T, Madorran E, Sala M, Barranco A (2017) Toxic effects of perfluorinated compounds at human cellular level and on a model vertebrate. Food Chem Toxicol 104:14–25. https://doi.org/10.1016/j.fct.2017.02.041

Reistad T, Fonnum F, Mariussen E (2013) Perfluoroalkylated compounds induce cell death and formation of reactive oxygen species in cultured cerebellar granule cells. Toxicol Lett 218:56–60. https://doi.org/10.1016/j.toxlet.2013.01.006

Rice PA (2015) C6-perfluorinated compounds: the new greaseproofing agents in food packaging. Curr Environ Health Rep 2:33–40. https://doi.org/10.1007/s40572-014-0039-3

Rice PA, Aungst J, Cooper J, Bandele O, Kabadi SV (2020) Comparative analysis of the toxicological databases for 6:2 fluorotelomer alcohol (6:2 FTOH) and perfluorohexanoic acid (PFHxA). Food Chem Toxicol 138:111210. https://doi.org/10.1016/j.fct.2020.111210

Ritter SK (2015) The shrinking case for fluorochemicals. Chem Eng News 93:27–29

Roberts LJ (1998) Measurement of lipid peroxidation AU – Moore, Kevin. Free Radic Res 28:659–671. https://doi.org/10.3109/10715769809065821

Rosenmai AK, Nielsen FK, Pedersen M, Hadrup N, Trier X, Christensen JH, Vinggaard AM (2013) Fluorochemicals used in food packaging inhibit male sex hormone synthesis. Toxicol Appl Pharm 266:132–142. https://doi.org/10.1016/j.taap.2012.10.022

Ruan T, Sulecki LM, Wolstenholme BW, Jiang G, Wang N, Buck RC (2014) 6:2 Fluorotelomer iodide in vitro metabolism by rat liver microsomes: Comparison with 1,2-C-14 6:2 fluorotelomer alcohol. Chemosphere 112:34–41. https://doi.org/10.1016/j.chemosphere.2014.02.068

Russell MH, Himmelstein MW, Buck RC (2015) Inhalation and oral toxicokinetics of 6:2 FTOH and its metabolites in mammals. Chemosphere 120:328–335. https://doi.org/10.1016/j.chemosphere.2014.07.092

Serex T, Anand S, Munley S, Donner EM, Frame SR, Buck RC, Loveless SE (2014) Toxicological evaluation of 6:2 fluorotelomer alcohol. Toxicology 319:1–9. https://doi.org/10.1016/j.tox.2014.01.009

Sha B, Dahlberg A-K, Wiberg K, Ahrens L (2018) Fluorotelomer alcohols (FTOHs), brominated flame retardants (BFRs), organophosphorus flame retardants (OPFRs) and cyclic volatile methylsiloxanes (cVMSs) in indoor air from occupational and home environments. Environ Pollut 241:319–330. https://doi.org/10.1016/j.envpol.2018.04.032

Shabalina IG, Panaretakis T, Bergstrand A, DePierre JW (1999) Effects of the rodent peroxisome proliferator and hepatocarcinogen, perfluorooctanoic acid, on apoptosis in human hepatoma HepG2 cells. Carcinogenesis 20:2237–2246. https://doi.org/10.1093/carcin/20.12.2237

Shi GH, Cui QQ, Pan YT, Sheng N, Guo Y, Dai JY (2017) 6:2 fluorotelomer carboxylic acid (6:2 FTCA) exposure induces developmental toxicity and inhibits the formation of erythrocytes during zebrafish embryogenesis. Aquat Toxicol 190:53–61. https://doi.org/10.1016/j.aquatox.2017.06.023

Shoeib T, Hassan Y, Rauert C, Harner T (2016) Poly- and perfluoroalkyl substances (PFASs) in indoor dust and food packaging materials in Egypt: Trends in developed and developing countries. Chemosphere 144:1573–1581. https://doi.org/10.1016/j.chemosphere.2015.08.066

Ssebugere P, Sillanpää M, Matovu H, Wang Z, Schramm K-W, Omwoma S (2020) Environmental levels and human body burdens of per- and poly-fluoroalkyl substances in Africa: a critical review. Sci Total Environ 739:139913. https://doi.org/10.1016/j.scitotenv.2020.139913

Takacs ML, Abbott BD (2007) Activation of mouse and human peroxisome proliferator–activated receptors (α, β/δ, γ) by perfluorooctanoic acid and perfluorooctane sulfonate. Toxicol Sci 95:108–117. https://doi.org/10.1093/toxsci/kfl135

Takagi A, Sai K, Umemura T, Hasegawa R, Kurokawa Y (1991) Short-term exposure to the peroxisome proliferators, perfluorooctanoic acid and perfluorodecanoic acid, causes significant increase of 8-hydroxydeoxyguanosine in liver DNA of rats. Cancer Lett 57:55–60. https://doi.org/10.1016/0304-3835(91)90063-N

Tang J, Jia X, Gao N, Wu Y, Liu Z, Lu X, Du Q, He J, Li N, Chen B, Jiang J, Liu W, Ding Y, Zhu W, Zhang H (2018a) Role of the Nrf2-ARE pathway in perfluorooctanoic acid (PFOA)-induced hepatotoxicity in Rana nigromaculata. Environ Pollut 238:1035–1043. https://doi.org/10.1016/j.envpol.2018.02.037

Tang J, Lu X, Chen F, Ye X, Zhou D, Yuan J, He J, Chen B, Shan X, Jiang J, Liu W, Zhang H (2018b) Effects of perfluorooctanoic acid on the associated genes expression of autophagy signaling pathway of *Carassius auratus* lymphocytes in vitro. Front Physiol 9:1748–1748. https://doi.org/10.3389/fphys.2018.01748

Tian M, Peng S, Martin FL, Zhang J, Liu L, Wang Z, Dong S, Shen H (2012) Perfluorooctanoic acid induces gene promoter hypermethylation of glutathione-S-transferase Pi in human liver L02 cells. Toxicology 296:48–55. https://doi.org/10.1016/j.tox.2012.03.003

Valentine JS, Doucette PA, Potter SZ (2005) Copper-Zinc superoxide dismutase and amyotrophic lateral sclerosis. Annu Rev Biochem 74:563–593. https://doi.org/10.1146/annurev.biochem.72.121801.161647

Wang N, Szostek B, Buck RC, Folsom PW, Sulecki LM, Capka V, Berti WR, Gannon JT (2005a) Fluorotelomer alcohol biodegradation – direct evidence that perfluorinated carbon chains breakdown. Environ Sci Technol 39:7516–7528

Wang N, Szostek B, Folsom PW, Sulecki LM, Capka V, Buck RC, Berti WR, Gannon JT (2005b) Aerobic biotransformation of C-14-labeled 8-2 telomer B alcohol by activated sludge from a domestic sewage treatment plant. Environ Sci Technol 39:531–538. https://doi.org/10.1021/es049466y

Wang N, Szostek B, Buck RC, Folsom PW, Sulecki LM, Gannon JT (2009) 8-2 fluorotelomer alcohol aerobic soil biodegradation: pathways, metabolites, and metabolite yields. Chemosphere 75:1089–1096. https://doi.org/10.1016/j.chemosphere.2009.01.033

Wang Y, Fu J, Wang T, Liang Y, Pan Y, Cai Y, Jiang G (2010) Distribution of perfluoroctane sulfonate and other perfluorochemicals in the ambient environment around a manufacturing facility in China. Environ Sci Technol 44:8062–8067

Wang X, Schuster J, Jones KC, Gong P (2018) Occurrence and spatial distribution of neutral perfluoroalkyl substances and cyclic volatile methylsiloxanes in the atmosphere of the Tibetan Plateau. Atmos Chem Phys 18:8745–8755. https://doi.org/10.5194/acp-18-8745-2018

Winkens K, Giovanoulis G, Koponen J, Vestergren R, Berger U, Karvonen AM, Pekkanen J, Kiviranta H, Cousins IT (2018) Perfluoroalkyl acids and their precursors in floor dust of children's bedrooms – implications for indoor exposure. Environ Int 119:493–502. https://doi.org/10.1016/j.envint.2018.06.009

Wolf CJ, Zehr RD, Schmid JE, Lau C, Abbott BD (2010) Developmental effects of perfluorononanoic acid in the mouse are dependent on peroxisome proliferator-activated receptor-alpha. PPAR Res. https://doi.org/10.1155/2010/282896

Wong F, Shoeib M, Katsoyiannis A, Eckhardt S, Stohl A, Bohlin-Nizzetto P, Li H, Fellin P, Su Y, Hung H (2018) Assessing temporal trends and source regions of per- and polyfluoroalkyl substances (PFASs) in air under the Arctic Monitoring and Assessment Programme (AMAP). Atmos Environ 172:65–73. https://doi.org/10.1016/j.atmosenv.2017.10.028

Xing J, Wang G, Zhao J, Wang E, Yin B, Fang D, Zhao J, Zhang H, Chen YQ, Chen W (2016) Toxicity assessment of perfluorooctane sulfonate using acute and subchronic male C57BL/6J mouse models. Environ Pollut 210:388–396. https://doi.org/10.1016/j.envpol.2015.12.008

Yang B, Zou WY, Hu ZZ, Liu FM, Zhou L, Yang SL, Kuang HB, Wu L, Wei J, Wang JL, Zou T, Zhang DL (2014a) Involvement of oxidative stress and inflammation in liver injury caused by perfluorooctanoic acid exposure in mice. Biomed Res Int. https://doi.org/10.1155/2014/409837

Yang S, Liu S, Ren Z, Jiao X, Qin S (2014b) Induction of oxidative stress and related transcriptional effects of perfluorononanoic acid using an in vivo assessment. Comp Biochem Phys C 160:60–65. https://doi.org/10.1016/j.cbpc.2013.11.007

Yao X, Zhong L (2005) Genotoxic risk and oxidative DNA damage in HepG2 cells exposed to perfluorooctanoic acid. Mutat Res-Gen Toxicol Environ 587:38–44. https://doi.org/10.1016/j.mrgentox.2005.07.010

Yu Y, Cui Y, Niedernhofer LJ, Wang Y (2016) Occurrence, biological consequences, and human health relevance of oxidative stress-induced DNA damage. Chem Res Toxicol 29:2008–2039. https://doi.org/10.1021/acs.chemrestox.6b00265

Yuan G, Peng H, Huang C, Hu J (2016) Ubiquitous occurrence of fluorotelomer alcohols in eco-friendly paper-made food-contact materials and their implication for human exposure. Environ Sci Technol 50(2):942–950

Zareitalabad P, Siemens J, Hamer M, Amelung W (2013) Perfluorooctanoic acid (PFOA) and perfluorooctanesulfonic acid (PFOS) in surface waters, sediments, soils and wastewater – a review on concentrations and distribution coefficients. Chemosphere. https://doi.org/10.1016/j.chemosphere.2013.02.024

Zelko IN, Mariani TJ, Folz RJ (2002) Superoxide dismutase multigene family: A comparison of the CuZn-SOD (SOD1), Mn-SOD (SOD2), and EC-SOD (SOD3) gene structures, evolution, and expression. Free Radical Bio Med 33:337–349. https://doi.org/10.1016/s0891-5849(02)00905-x

Zhang L, Niu J, Li Y, Wang Y, Sun D (2013a) Evaluating the sub-lethal toxicity of PFOS and PFOA using rotifer Brachionus calyciflorus. Environ Pollut 180:34–40. https://doi.org/10.1016/j.envpol.2013.04.031

Zhang S, Szostek B, McCausland PK, Wolstenholme BW, Lu X, Wang N, Buck RC (2013b) 6:2 and 8:2 fluorotelomer alcohol anaerobic biotransformation in digester sludge from a WWTP under methanogenic conditions. Environ Sci Technol 47:4227–4235. https://doi.org/10.1021/es4000824

Zhang S, Merino N, Wang N, Ruan T, Lu X (2017a) Impact of 6:2 fluorotelomer alcohol aerobic biotransformation on a sediment microbial community. Sci Total Environ 575:1361–1368. https://doi.org/10.1016/j.scitotenv.2016.09.214

Zhang S, Merino N, Wang N, Ruan T, Lu XX (2017b) Impact of 6:2 fluorotelomer alcohol aerobic biotransformation on a sediment microbial community. Sci Total Environ 575:1361–1368. https://doi.org/10.1016/j.scitotenv.2016.09.214

Zhang J, Wang B, Zhao B, Li Y, Zhao X, Yuan Z (2019) Blueberry anthocyanin alleviate perfluorooctanoic acid-induced toxicity in planarian (Dugesia japonica) by regulating oxidative stress biomarkers, ATP contents, DNA methylation and mRNA expression. Environ Pollut 245:957–964. https://doi.org/10.1016/j.envpol.2018.11.094

Zhao GP, Wang J, Wang XF, Chen SR, Zhao Y, Gu F, Xu A, Wu LJ (2011) Mutagenicity of PFOA in mammalian cells: role of mitochondria-dependent reactive oxygen species. Environ Sci Technol 45:1638–1644. https://doi.org/10.1021/es1026129

Zhao M, Jiang QX, Wang WC, Geng M, Wang M, Han YT, Wang CB (2017a) The roles of reactive oxygen species and nitric oxide in perfluorooctanoic acid-induced developmental cardiotoxicity and L-carnitine mediated protection. Int J Mol Sci 18. https://doi.org/10.3390/ijms18061229

Zhao Z, Tang J, Mi L, Tian C, Zhong G, Zhang G, Wang S, Li Q, Ebinghaus R, Xie Z, Sun H (2017b) Perfluoroalkyl and polyfluoroalkyl substances in the lower atmosphere and surface waters of the Chinese Bohai Sea, Yellow Sea, and Yangtze River estuary. Sci Total Environ 599:114–123. https://doi.org/10.1016/j.scitotenv.2017.04.147

Perchlorate Contamination: Sources, Effects, and Technologies for Remediation

Rosa Acevedo-Barrios and Jesus Olivero-Verbel

Contents

1 Introduction 104
2 Origins and Utilisation of Perchlorate 105
   2.1 Natural Origin 105
   2.2 Anthropogenic Origin 106
3 Environmental and Health Effects 107
   3.1 Health Effects on Humans 107
   3.2 Effects on Biota 107
4 Technologies for Perchlorate Treatment 108
   4.1 Physicochemical Methods 108
   4.2 Biological Treatment of Perchlorate 110
5 Conclusions 115
References 115

Abstract Perchlorate is a persistent pollutant, generated via natural and anthropogenic processes, that possesses a high potential for endocrine disruption in humans and biota. It inhibits iodine fixation, a major reason for eliminating this pollutant from ecosystems. Remediation of perchlorate can be achieved with various physicochemical treatments, especially at low concentrations. However, microbiological approaches using microorganisms, such as those from the genera *Dechloromonas*, *Serratia*, *Propionivibrio*, *Wolinella*, and *Azospirillum*, are promising when perchlorate pollution is extensive. Perchlorate-reducing bacteria, isolated from harsh

R. Acevedo-Barrios
Environmental and Computational Chemistry Group, School of Pharmaceutical Sciences, University of Cartagena, Cartagena, Colombia

Grupo de Investigación en Estudios Químicos y Biológicos, Facultad de Ciencias Básicas, Universidad Tecnológica de Bolívar, Cartagena, Colombia
e-mail: racevedo@utb.edu.co

J. Olivero-Verbel (✉)
Environmental and Computational Chemistry Group, School of Pharmaceutical Sciences, University of Cartagena, Cartagena, Colombia
e-mail: joliverov@unicartagena.edu.co

© The Author(s), under exclusive license to Springer Nature Switzerland AG 2021
P. de Voogt (ed.), *Reviews of Environmental Contamination and Toxicology Volume 256*, Reviews of Environmental Contamination and Toxicology 256,
https://doi.org/10.1007/398_2021_66

environments, for example saline soils, mine sediments, thermal waters, wastewater treatment plants, underground gas storage facilities, and remote areas, including the Antarctica, can provide removal yields from 20 to 100%. Perchlorate reduction, carried out by a series of enzymes, such as perchlorate reductase and superoxide chlorite, depends on pH, temperature, salt concentration, metabolic inhibitors, nutritional conditions, time of contact, and cellular concentration. Microbial degradation is cost-effective, simple to implement, and environmentally friendly, rendering it a viable method for alleviating perchlorate pollution in the environment.

Keywords Bacteria · Biological treatment · Environmental pollutant · Perchlorate-reducing · Toxicology

Highlights

- Perchlorate is a contaminant generated via natural and anthropogenic processes.
- It is persistent in the environment and exerts endocrine effects on humans and biota.
- Physicochemical processes are useful for treatment of low perchlorate concentrations.
- Bacteria-mediated remediation is suitable for treating perchlorate-polluted sites.
- *Dechloromonas*, *Serratia*, and *Propionivibrio* are promising perchlorate degradators.

1 Introduction

Perchlorate, a chemically stable anion, is a powerful oxidiser (Cao et al. 2019). It possesses a tetrahedral structure containing a chlorine atom surrounded by four oxygen atoms (Murray and Bolger 2014). Perchlorate salts are water soluble, being rapidly incorporated into aquatic ecosystems from polluted soils by runoff-related processes. Once in the water bodies, it bio-accumulates in leafy vegetables (Urbansky 2002). Animals, however, show lower perchlorate concentrations in their tissues, compared to those found in their environment, likely as a result of its hydrophilic nature (Lee et al. 2012) that allows its elimination by the urine.

Perchlorate has recently become a major inorganic contaminant in drinking water; therefore, the EPA has established an official reference dose (RfD) of 0.0007 mg/kg/day perchlorate (EPA 2005; Srinivasan and Sorial 2009).

Perchlorate is considered an endocrine disruptor that affects the thyroid function by inhibiting iodine uptake (Pleus and Corey 2018; Bardiya and Bae 2011). That is the main reason it is necessary to identify its sources, environmental and health effects, as well as to eliminate this pollutant from ecosystems (Kumarathilaka et al. 2016).

Because of the stable chemical structure of perchlorate, using reducing physico-chemical agents is effective especially in the treatment of low concentrations (Srinivasan and Sorial 2009), but it is ineffective for transforming it into less toxic forms (Ghosh et al. 2014). Microbial perchlorate-reducing technologies may potentially aid in the reduction and degradation of this pollutant (Bardiya and Bae 2011; Wang et al. 2014).

This review provides an elaborate discussion of recent issues regarding perchlorate remediation, attempting to include different perspectives from recent research, in particular the origins and use of perchlorate, its environmental effects on human and biota health, and the various treatment technologies for the control of perchlorate contamination.

2 Origins and Utilisation of Perchlorate

The presence of perchlorate in the environment stems from human activities and natural sources. Drinking water is likely the greatest source of ClO_4^- exposure (Steinmaus 2016; Cao et al. 2019).

Perchlorate enters the body via trophic transfer because it is taken up into plants and crops from contaminated soils and irrigation waters. This is considered one of the most important routes of exposure. For these reasons, the presence of perchlorate in foods has been extensively studied and documented (Zhang et al. 2010; Lee et al. 2012). Perchlorate is present in certain crops such as lettuce, carrots, rice, spinach, and fruits (USFDA 2008; Calderón et al. 2017). Perchlorate is also detected in the milk of cattle, as well as in sausages, ham, instant noodles, fish, meat (Okeke et al. 2002; USFDA 2008; Lee et al. 2012; Maffini et al. 2016), tea, sodas, and tobacco plants and products (ATSDR 2008). Moreover, recent studies have detected perchlorate in breast milk (Wang et al. 2008a; Ye et al. 2012).

Other common routes of exposure include the ingestion and inhalation of domestic dust or atmospheric depositions can act as carriers of contaminants, affecting the health of children and adults. The daily ingested dose of perchlorate can reach 0.7 µg/kg/day for powdered perchlorate, 0.86 ng/mL for perchlorate in tap water, 1.03 ng/mL for perchlorate in bottled water, 536 ng/g for perchlorate in food, 160 ng/mL for inhaled perchlorate that is present in the atmosphere, and 0.03 ng/kg for dermal exposure (Gan et al. 2014; Wan et al. 2015; Kumarathilaka et al. 2016).

2.1 Natural Origin

Perchlorate formation can occur naturally in the environment. Recent findings indicate that perchlorate is continuously formed in the atmosphere by chlorine or sodium chloride reacting with ozone, likely via photochemical processes or electric activity during storms (USFDA 2008; Dasgupta et al. 2005). Other possible sources are volcano eruptions (Simonaitis and Heicklen 1975; Furdui et al. 2018). High concentrations of perchlorate are found in arid (Murray and Bolger 2014) and/or hypersaline environments (Ryu et al. 2012; Acevedo-Barrios et al. 2019).

The most abundant perchlorate deposits are found in arid regions of South America, particularly in Chile, Peru, and Bolivia, especially in soils from arid and semi-arid regions, where it is usually mixed with calcium carbonate, gypsum, and sodium salts (Murray and Bolger 2014). Deposits containing perchlorate have also been found in Death Valley in the USA (Jackson et al. 2010). These are also rich in sodium and potassium nitrate, being extracted and exported worldwide for nearly 200 years, still generating increased interest (Murray and Bolger 2014).

Interestingly, perchlorate salts have also been found in kelp forests and the Antarctica (Jackson et al. 2010, 2015; Kounaves et al. 2010; Kumarathilaka et al. 2016). The perchlorate and chlorate detected in Antarctic lakes were determined to be of atmospheric origin, where perchlorate concentration is approximately 1,100 µg/kg.

It is known that perchlorate is a pollutant in mined nitrate salts that are used as fertilisers, being this a common pathway to reach soils used for agriculture purposes (Urbansky et al. 2001; Aziz et al. 2006; Sanchez et al. 2006). However, little is known about perchlorate levels in soils fertilised with nitrates and their impacts on agricultural systems and food security (Calderon et al. 2020).

2.2 Anthropogenic Origin

Most of the perchlorate distributed in environmental matrices results from anthropogenic activities. Perchlorate contamination has been increasingly generated by anthropogenic sources, with concentrations of 0.1–35.0 µg/L in drinking water (Blount et al. 2010), 0.1–22.1 µg/L in groundwater, 1.0–2,300 µg/L in surface water (Wu et al. 2010), and 1.0–13 µg/kg in soil (Jackson et al. 2010; Ye et al. 2013). In groundwater, perchlorate contamination may result from rainwater percolating through contaminated sand or soil (ATSDR 2008).

A substantial portion of perchlorate found in ground and surface water is associated with the military industry (Hubé and Urban 2013; Cao et al. 2019), aerospace industry, and manufacture of explosives (Murray and Bolger 2014; Kumarathilaka et al. 2016).

Fireworks, which contain chemicals such as potassium nitrate, potassium chloride, potassium or ammonium perchlorate, coal, sulphur, manganese, sodium, nitrate, strontium, aluminium, barium titanate, and powdered iron oxalate, are one of the main sources of perchlorate pollution in the environment (Wu et al. 2011; Vellanki et al. 2013; Ye et al. 2013).

Perchlorate is found in polyvinyl chloride (PVC) and in Li-ion batteries, in which it is used as dopant material (Interstate Technology Regulatory Council 2005); this indicates that residents and workers at electronic waste-recycling sites are likely exposed to this pollutant. Additionally, perchlorate is a strong oxidiser and is used as a cleaning agent during the production of LCD screens (Her et al. 2011).

Ammonium, lithium, magnesium, and potassium perchlorates are used during manufacture of herbicides and automotive airbag inflators. Potassium perchlorate has been used clinically in 1950–1960 to treat hyperthyroidism (Murray and Bolger 2014). Perchlorate is also utilised in the manufacture of matches, dyes, rubber, and lubricants (Interstate Technology Regulatory Council 2005). Other anthropogenic sources of perchlorate include foundries, road flares, drying and engraving agents, gunpowder, batteries (Wang et al. 2014), disinfectants, bleach, chlorine-based cleaners and chemicals for swimming pool chlorination, electronic tubes, paints, enamels, fertilisers, nuclear reactors, and other materials for commercial use (Interstate Technology Regulatory Council 2005; Agency for Toxic Substances and Disease Registry 2008; Kumarathilaka et al. 2016; Maffini et al. 2016). Additionally, perchlorate is a growth promoter and has been added to thyrostatic medication used for the fattening of livestock (Gholamian et al. 2011).

3 Environmental and Health Effects

3.1 Health Effects on Humans

Perchlorate affects the normal function of the thyroid gland (Murray and Bolger 2014). It inhibits iodine fixation, a process necessary for the production of thyroid hormones, leading to hypothyroidism (Ghosh et al. 2014; Murray and Bolger 2014; Cao et al. 2019). The thyroid gland, considered a major target for perchlorate, plays an important role in the regulation of metabolism and is essential for normal growth and development in children (Bruce et al. 1999; Agency for Toxic Substances and Disease Registry 2008; Bardiya and Bae 2011; Murray and Bolger 2014; Zhang et al. 2016).

Perchlorate can also induce damage to the nervous, reproductive, and immune systems (Gholamian et al. 2011), as well as induce teratogenesis in pregnant women, where it has been found in breast milk, saliva, and urine (Thrash et al. 2007; Zhang et al. 2010). Because it is found in breast milk, perchlorate is dangerous to nursing mothers and infants. Once ingested by the infant, it can impair thyroid function and cause physical and mental disabilities (Morreale de Escobar et al. 2000).

3.2 Effects on Biota

One study has shown that perchlorate is easily accumulated in plants (Andraski et al. 2014). Xie et al. (2014) showed that the rice plant *Oryza sativa* L. is easily contaminated by perchlorate and suggested that perchlorate may inhibit the growth of the plant. Perchlorate also affects chlorophyll content and root systems of *Acorus calamus, Canna indica, Thalia dealbata,* and *Eichhornia crassipes* (He et al. 2013). The study by Acevedo-Barrios et al. (2018) has shown that the survival of freshwater algae *Pseudokirchneriella subcapitata* is considerably impaired by perchlorate ($LC_{50} = 72$ mM). However, the exact manner by which perchlorate damages the photosystem is unclear, and further studies are needed to understand its mechanisms of action (Xie et al. 2014).

Environmental perchlorate concentrations of 200–500 µg/L disrupt metamorphosis in amphibians (Goleman et al. 2002), and concentrations ≥ 500 µg/L cause diverse fish alterations (Schmidt et al. 2012). Perchlorate decreases the number of eggs spawned by the Japanese medaka (*Oryzias latipes*) (Lee et al. 2014), modifies sex ratio and decreases the growth rates of zebrafish (Liu et al. 2008; Schmidt et al. 2012), and causes functional hermaphroditism in *Gasterosteus aculeatus* females. These findings indicate that perchlorate exerts androgenic effects. Other effects of perchlorate on fish include abnormal development of the lateral plates, decreased swimming performance, slow growth rates, and reduced pigmentation (Bernhardt et al. 2011).

In *V. fischeri*, perchlorate-induced toxicity manifests as reduced bioluminescence, whereas in the crustaceans *D. magna* and *E. fetida,* perchlorate exposure disrupts endocrine function and can result in mortality (Acevedo-Barrios et al. 2019). In *E. fetida*, perchlorate can also induce avoidance behaviour, weight loss, decreased production of eggs and hatchlings, malformations, dwarfism, and necrosis. These findings demonstrate that perchlorate toxicity varies according to species (Acevedo-Barrios et al. 2018).

Exposure to perchlorate leads to embryonic hypothyroidism in bird species, such as the Japanese quail (Chen et al. 2008), and affects thyroid function in species such as *Colinus virginianus* and *Anas platyrhynchos* (McNabb 2003). Similarly, perchlorate affects reptiles and mammals such as lizards, mice, rats, and rabbits (USCHPPM 2007).

4 Technologies for Perchlorate Treatment

Perchlorate present in soils contaminates surface and groundwater (Kumarathilaka et al. 2016) and may threaten the ecosystem and human health. Some of the strategies used to remove perchlorate from environment are described below.

4.1 Physicochemical Methods

These reactions are divided into two types: sequestration and transformation. Sequestration reactions include different types of membrane-based separation, ion exchange, and precipitation and transformation reactions include chemical reduction, electrochemistry-based and activated carbon approaches (Coates and Jackson 2009; Ghosh et al. 2014).

Physicochemical methods are promising treatments, especially in the treatment of low concentrations of perchlorate (Coates and Jackson 2009).

4.1.1 Membrane-Based Technologies

Membrane filtration, via reverse osmosis, nanofiltration, and ultrafiltration, has been somewhat effective for perchlorate removal (Ye et al. 2012). Membrane-based techniques can be effective; however, membrane incrustations and high costs present drawbacks to these approaches (Srinivasan and Sorial 2009). Moreover, the rate of perchlorate diffusion decreases with increasing pH due to membrane surface charge becoming more negative (Yoon et al. 2005); and although membrane-based systems can remove various compounds, those can leave residues, requiring additional treatment (Coates and Jackson 2009).

In addition, water treated using this system has to be remineralised with sodium chloride, sodium bicarbonate, and other harmless salts to prevent degradation of the distribution system and render the water appealing to the consumer (Ghosh et al. 2014; Kumarathilaka et al. 2016).

4.1.2 Ion Exchange

Ion exchange is a promising technology for perchlorate elimination, offering a wide variety of strong anion exchange resins highly selective for perchlorate (Chen et al. 2012; Srinivasan and Sorial 2009; Batista et al. 2000). The perchlorate-laden spent resins require regeneration, which results in the production of concentrated brine (Bardiya and Bae 2011) that requires subsequent treatment (Srinivasan and Sorial 2009; Ghosh et al. 2011). The resins utilised in this approach are single-use, rendering this technology incomplete and economically unsustainable for perchlorate elimination (Ye et al. 2012).

The efficiency of this treatment primarily depends on the type of ion exchange matrix and concentration of other ions in the water (Coates and Jackson 2009). If concentration of other ions is higher than that of perchlorate, these other ions can compete for available sites on the resin, thereby reducing the capacity for perchlorate binding (Srinivasan and Sorial 2009).

During water treatment, this technology can produce salt resins in the presence of other anions, which constitutes a drawback. Therefore, for ion exchange, it is necessary to demineralise or remineralise the water being treated, depending on its anion content (Ghosh et al. 2014; Kumarathilaka et al. 2016).

4.1.3 Chemical Reduction

Chemical perchlorate reduction has been studied extensively because numerous metals can reduce perchlorate to chloride. For example, Hurley and Shapley

(2007) developed a bimetal Pd/Rh catalyst that exerts a rapid reduction of perchlorate in the presence of hydrogen, but it showed two limitations: the reaction occurred at a pH less than 3, and requires a pressure of 5 bars (Srinivasan and Sorial 2009). Similarly, Wang et al. (2008b) eliminated 90% of perchlorate ions over 3 days, using hydrogen gas and metallic Ti-TiO$_2$ catalysts in a pressurised reactor (Srinivasan and Sorial 2009). Despite high conversion yields, the use of this technology in environmental systems is not efficient due to the slow speed of the reaction process, costs, and technical factors (Coates and Jackson 2009; Ghosh et al. 2014; Kumarathilaka et al. 2016). Moreover, chemical reductions can be effective for decontamination of small systems, but cannot be scaled to real conditions in large water-treatment plants (Srinivasan and Sorial 2009).

4.1.4 Electrochemical Reduction

Electrochemical reduction is a promising removal technology because it completely destroys perchlorate without the use of catalysts; however, this approach can be implemented on a large scale (Srinivasan and Sorial 2009).

Historically, perchlorate has been used as an inert electrolyte in corrosion and electrochemical studies. Removing this pollutant through electrochemical approaches is a slow process (Srinivasan and Sorial 2009), rendering them ineffective for perchlorate reduction. Moreover, under environmental conditions, the reactions involved require large surfaces and can be affected by the presence of other reactive and non-reactive species. Although electrochemical processes are not suitable for *in situ* applications, those can be used to treat concentrated solutions (Coates and Jackson 2009; Kumarathilaka et al. 2016; Theis et al. 2002).

4.1.5 Activated Carbon

Granular activated carbon adsorption is simple to retrofit for targeting perchlorate in water and is, therefore, widely used for the treatment of drinking water (Srinivasan and Sorial 2009). This technique, however, is expensive because it has limited capacity for perchlorate adsorption and generates brines that can affect the effectiveness of the process (Na et al. 2002; Parette et al. 2005; Srinivasan and Sorial 2009).

4.2 *Biological Treatment of Perchlorate*

This technology implements biological systems (such as bacteria, algae, fungi, yeast, and plants) to remove or recover the pollutant from environmental matrices.

4.2.1 Types of Perchlorate-Reducing Bacteria

Perchlorate-reducing bacteria are phylogenetically diverse and include classes such as Alphaproteobacteria, Betaproteobacteria, Gammaproteobacteria, and Deltaproteobacteria, with Betaproteobacteria being the most commonly detected (Wallace et al. 1996; Coates et al. 1999; Bruce et al. 1999; Waller et al. 2004; Ye et al. 2012; Acevedo-Barrios et al. 2019). Although many details remain unknown, the enzymes perchlorate reductase and chlorite dismutase have been marked as critical for the reduction or elimination of perchlorate into Cl^- and O_2 (Youngblut et al. 2016; Xu and Logan 2003).

Table 1 illustrates the variety of currently known perchlorate-reducing species.

Table 1 Genera and species of currently known perchlorate-reducing bacteria

Bacterial genus and species	Percentage of reduction (%)	Environmental conditions and observations	Reference
Nesiotobacter sp.	25	FC, 37°C	Acevedo-Barrios et al. (2019)
Bacillus vallimostis	23	FC, 37°C	Acevedo-Barrios et al. (2019)
Salinivibrio costicola	25	FC, 37°C	Acevedo-Barrios et al. (2019)
Vibrio	14	FC, 37°C	Acevedo-Barrios et al. (2019)
Bacillus	12	FC, 37°C	Acevedo-Barrios et al. (2019)
Staphylococcus	10	FC, 37°C	Acevedo-Barrios et al. (2019)
Alteromonadaceae	30	ST, in presence of nitrate	Stepanov et al. (2014)
Azospirillum sp.	100	AN, 22°C	Waller et al. (2004)
Citrobacter sp.	32	30°C, pH 7, ST (0–5%)	Rikken et al. (1996); Okeke et al. (2002); Wang et al. (2008a)
Clostridium sp.	80	FC, 37°C	Chung et al. (2009)
Dechlorobacter hydrogenophilus	100	37°C, pH 6.5	Thrash et al. (2007)
Dechloromonas agitata	100	AN	Bruce et al. (1999); Sun et al. (2009); Vigliotta et al. (2010)
Dechloromarinus cepa NSS	30	37–42°C, pH 7.5, ST (2.5%)	Bardiya and Bae (2011)
Dechlorospirillum sp.	70	AN, 37°C	Coates et al. (1999); Vigliotta et al. (2010)
Escherichia coli K-12 cepa NAR gen G	90	AN, 37°C	Vigliotta et al. (2010)

(continued)

Table 1 (continued)

Bacterial genus and species	Percentage of reduction (%)	Environmental conditions and observations	Reference
Halomonas halodenitrificans	20	ST (5%)	Logan et al. (2001); Okeke et al. (2002)
Ideonella Dechloratans	90	AN 30°C	Bruce et al. (1999); Lindqvist et al. (2012)
Magnetospirillum magnetotacticum	70	AE, 37°C	Xu and Logan (2003)
Marinobacter sp.	7–11	ST (5%)	Ahn et al. (2009); Stepanov et al. (2014)
Propionivibrio militaris	100	AE, 30°C, ST (5%), pH 6.8	Gu and Brown (2006)
Proteus mirabilis	90	FC, AN, 30°C	Bruce et al. (1999)
Rhodobacter capsulatus	80	FC, 30°C	Bruce et al. (1999)
Serratia Marcescens	100	ST (>15%); pH 4.0 to 9.0	Vijaya Nadaraja et al. (2013); Sankar et al. (2014)
Vibrio dechloraticans	20	FC, AN. Degradation in concentrations of 1, 5, 25, 50, 75, 100, 125, 150, 175, 200 µg/L	Wang et al. (2008a)
Wolinella sp.	99.9	FC, AE, in presence of hydrogen	Rikken et al. (1996)

ST salt-tolerant (percentages of NaCl), *FC* facultative conditions, *AN* anaerobic conditions

4.2.2 Habitats of Perchlorate-Reducing Bacteria

Phenotypic characterisation studies have shown that perchlorate-reducing bacteria exhibit a wide range of metabolic capabilities and can thrive in diverse and extreme environments such as saline lakes, hot springs, and even hyperthermophilic and hypersaline soils (Rikken et al. 1996; Wallace et al. 1996; Coates et al. 1999; Bruce et al. 1999; Logan et al. 2001; Jackson et al. 2012; Matsubara et al. 2016; Acevedo-Barrios et al. 2019). In hypersaline soils, perchlorate is found at 25–2,700 mg/kg; these depositions are due to a shortage of rainfall. Marine soils usually contain bacteria species with biochemical versatility and ability to tolerate salt, being an interesting target for researchers due to the potential reduction of environmental perchlorate (Logan et al. 2001). The reason for selecting this type of environment is that degradation of perchlorate may be carried out using salt-tolerant bacteria (Okeke et al. 2002), although this perchlorate-reduction process could be impaired with increasing salinity (Vijaya Nadaraja et al. 2013; Matsubara et al. 2016).

Currently, no single technology can completely remove perchlorate from drinking water. However, combining the technologies described in this review may be a feasible approach (Srinivasan and Sorial 2009).

For example, combining physicochemical treatments and perchlorate-reducing halophilic bacteria may increase the efficiency of perchlorate reduction, allow the treatment of matrices with high salinity and perchlorate concentrations, and resolve the persistent issue of waste disposal (Xiao and Roberts 2013).

4.2.3 Environmental Requirements for Bacterial Degradation of Perchlorate

In biosorption, the capacity of perchlorate sorption by the microbial system is regulated by several factors. These include environmental factors, such as pH, temperature, salt concentration, presence of metabolic inhibitors or electron acceptors, nutritional conditions, and time of contact; and physiological factors, including type and physiological age of the microorganism, biomass condition, cellular concentration, and mutations in the bacteria being used for reduction (Coates et al. 1999; Logan et al. 2001; Ting et al. 2008). Perchlorate degradation is generally much slower in the pH range of 5.0–9.0 (Wang et al. 2008a; Wan et al. 2016; Zhu et al. 2016).

Reduction is generally inhibited by the presence of nitrate (Coates and Achenbach 2004; Wan et al. 2016) because some reducing microorganisms prefer other electron acceptors over perchlorate (Coates and Jackson 2009). To prevent this, donor species are added in excess to remove non-perchlorate electron acceptors before performing the reduction; this is done because non-perchlorate electron acceptors can activate bacteria that do not degrade perchlorate, which results in inefficient treatment. Oxygen is another inhibitor of microbial reduction of perchlorate because its presence can cause bacteria to utilise donors for oxygen consumption (Coates and Jackson 2009; Xu et al. 2015). Studies have shown that perchlorate reduction should ideally be performed under anaerobic facultative conditions (Shrout et al. 2005; Acevedo-Barrios et al. 2019).

Generally, perchlorate reduction occurs at a temperature range of 10–40°C, but an ideal condition requires temperatures between 28 and 37°C (Coates et al. 1999; Zhu et al. 2016; Acevedo-Barrios et al. 2019). A low salt content (<2% NaCl) in an environment can be a limiting factor. In contrast, reductions have been performed in matrices having salt concentrations higher than 11% (Logan et al. 2001; Okeke et al. 2002).

4.2.4 Biochemical Metabolism of Perchlorate-Reducing Bacteria

Perchlorate-reducing bacteria are omnipresent and easily obtainable from most environments (Coates et al. 1999). These organisms, which show a wide range of metabolic activities (Chaudhuri et al. 2002), contain proteins that can degrade contaminated matrices on-site; unlike other methods, this approach causes minimal physical disturbance around the treated area (Volesky 1999). For these reasons, perchlorate-degrading bacteria can be used for various applications, including biotechnological processes (Wang et al. 2014), and represent a promising, effective, and economically viable solution for resolving perchlorate contamination (Logan et al. 2001; Acevedo-Barrios et al. 2016, 2019). Perchlorate-degrading bacteria utilise two key enzymes that reduce the activation energy required for perchlorate reduction and use perchlorate as an electron acceptor in their metabolic reactions (Coates et al. 2000; Jackson et al. 2015).

Enzymes, such as perchlorate reductase and superoxide chlorite, carry out the reduction or elimination of perchlorate. A reductase can reduce perchlorate to chlorate, and subsequently, to chlorite, whereas superoxide chlorite changes chlorite to chloride and molecular oxygen. Biological reduction of perchlorate using bacteria completely degrades perchlorate ions into Cl^- and O_2, as shown in Eq. 1 (Xu and Logan 2003; Matsubara et al. 2016; Acevedo-Barrios et al. 2019):

$$\left(ClO_4^-\right) (\text{Perchlorate}) \rightarrow \left(ClO_3^-\right) (\text{Chlorate}) \rightarrow \left(ClO_2^-\right) (\text{Chlorite}) \rightarrow Cl^- (\text{Chloride}) + O_2 \tag{1}$$

Equation 1: Perchlorate-degradation pathway

Perchlorate-reducing bacteria are facultative anaerobes or microaerophilic because molecular oxygen is produced as an intermediate of the microbial perchlorate reduction (Rikken et al. 1996; Wallace et al. 1996; Bruce et al. 1999; Coates and Achenbach 2006; Jackson et al. 2015).

4.2.5 Application of Biological Reduction

Biotransformation is a method that contributes to environmental sanitation and involves the removal of perchlorate via microorganisms that selectively retain ions found in the solution. Consequently, these microorganisms either deteriorate these ions into less toxic forms or remove them completely (Logan et al. 2001).

This technology is used for bioremediation in and ex situ and for natural biodegradation. The ex-situ treatment process is suitable for waste that has high concentrations of perchlorate, such as water from facilities that manufacture ammunition. Bioremediation in situ is suitable for reducing concentrations of perchlorate in shallow (<15 m deep) or narrow zones such as groundwater. This technology is cost-effective with respect to transportation of materials, poses low risk for accidents, and does not require bioaugmentation. This approach, which can be used to treat unsaturated zones and soils, stimulates native microflora via the addition of carbon sources and electron donors (Bardiya and Bae 2011); these factors render this method less costly than ex-situ bioremediation (Ye et al. 2012). The use of heterotrophic and autotrophic system for perchlorate removal has also been proposed (Li et al. 2019), combining sulphuric-based autotrophic processes to increase performance and reduce high concentrations of perchlorate.

5 Conclusions

Perchlorate is a contaminant that is generated via natural and anthropogenic processes naturally formed in the atmosphere and likely released during volcanic eruptions (Simonaitis and Heicklen 1975). This compound is persistent in the environment and exerts endocrine effects on humans and biota, affecting thyroid function, growth, and reproduction (Gholamian et al. 2011; Acevedo-Barrios et al. 2018). Physicochemical methods are used to remove perchlorate, but these techniques can only separate this contaminant from matrices, producing residues in the process (Coates and Jackson 2009), but is effective especially in the treatment of low concentrations.

Because perchlorate is kinetically stable and inert at low concentrations, most traditional physicochemical processes are not applicable for the elimination and decomposition of perchlorate ion (Logan et al. 2001). These methods are also costly to maintain and operate, have high energy requirements, and generate excessive amount of brines and resins with high concentrations of the pollutant. Using bacteria that can reduce and eliminate perchlorate is an effective and economically feasible approach (Hatzinger 2005; Acevedo-Barrios et al. 2019).

Among such bacteria, organisms most common at perchlorate reduction belong to the genera *Dechloromonas*, *Serratia*, *Propionivibrio*, *Wolinella*, and *Azospirillum*. These organisms are mostly isolated from adverse environments and degrade 20–100% of perchlorate; their removal efficiency depends on pH, temperature, salt concentration, presence of metabolic inhibitors, nutritional conditions, time of contact, cellular concentration, and other factors, indicating that bacteria-mediated remediation of perchlorate is a suitable method for controlling this type of contamination.

Research is currently underway to find a novel technology to remediate perchlorate, and although microbial reduction and ion exchange technologies are used, a combination of several technologies is required to remove this pollutant from ecosystems (Srinivasan and Sorial 2009).

Acknowledgements The authors thank the Research Department at the Technological University of Bolivar, and the Plan to Support Research Groups and Doctoral Programs at the University of Cartagena.

Conflict of Interest The authors have no conflict of interest to declare.

References

Acevedo-Barrios R, Bertel-Sevilla A, Alonso-Molina J, Olivero-Verbel J (2016) Perchlorate tolerant bacteria from saline environments at the Caribbean region of Colombia. Toxicol Lett: S103

Acevedo-Barrios R, Sabater-Marco C, Olivero-Verbel J (2018) Ecotoxicological assessment of perchlorate using in vitro and in vivo assays. Environ Sci Pollut Res 25:13697–13708. https://doi.org/10.1007/s11356-018-1565-6

Acevedo-Barrios R, Bertel-Sevilla A, Alonso-Molina J, Olivero-Verbel J (2019) Perchlorate-reducing bacteria from hypersaline soils of the Colombian Caribbean. Int J Microbiol 2019:1–13. https://doi.org/10.1155/2019/6981865

Agency for Toxic Substances and Disease Registry (2008) Toxicological profile for perchlorates

Ahn CH, Oh H, Ki D et al (2009) Bacterial biofilm-community selection during autohydrogenotrophic reduction of nitrate and perchlorate in ion-exchange brine. Appl Microbiol Biotechnol 81:1169–1177. https://doi.org/10.1007/s00253-008-1797-3

Andraski BJ, Jackson WA, Welborn TL et al (2014) Soil, plant, and terrain effects on natural perchlorate distribution in a desert landscape. J Environ Qual 43:980–994. https://doi.org/10.2134/jeq2013.11.0453

Aziz C, Borch R, Nicholson P, Cox E (2006) Alternative causes of wide-spread, low concentration perchlorate impacts to groundwater. In: Gu B, Coates J (eds) Perchlorate. Springer, pp 71–91

Bardiya N, Bae JH (2011) Dissimilatory perchlorate reduction: a review. Microbiol Res 166:237–254. https://doi.org/10.1016/j.micres.2010.11.005

Batista JR, McGarvey FX, Vieira AR (2000) The removal of perchlorate from waters using ion-exchange resins. In: Urbansky ET (ed) Perchlorate in the environment. Springer, Boston, pp 135–145

Bernhardt RR, Von Hippel FA, O'Hara TM (2011) Chronic perchlorate exposure causes morphological abnormalities in developing stickleback. Environ Toxicol Chem 30:1468–1478

Blount BC, Alwis KU, Jain RB et al (2010) Perchlorate, nitrate, and iodide intake through tap water. Environ Sci Technol 44:9564–9570

Bruce RA, Achenbach LA, Coates JD (1999) Reduction of (per)chlorate by a novel organism isolated from paper mill waste. Environ Microbiol 1:319–329. https://doi.org/10.1046/j.1462-2920.1999.00042.x

Calderón R, Godoy F, Escudey M, Palma P (2017) A review of perchlorate (ClO_4^-) occurrence in fruits and vegetables. Environ Monit Assess 189(2):82. https://doi.org/10.1007/s10661-017-5793-x

Calderon R, Rajendiran K, Kim UJ et al (2020) Sources and fates of perchlorate in soils in Chile: a case study of perchlorate dynamics in soil-crop systems using lettuce (Lactuca sativa) fields. Environ Pollut 264:114682. https://doi.org/10.1016/j.envpol.2020.114682

Cao F, Jaunat J, Sturchio N et al (2019) Worldwide occurrence and origin of perchlorate ion in waters: a review. Sci Total Environ 661:737–749. https://doi.org/10.1016/j.scitotenv.2019.01.107

Chaudhuri SK, O'Connor SM, Gustavson RL et al (2002) Environmental factors that control microbial perchlorate reduction. Appl Environ Microbiol 68:4425–4430. https://doi.org/10.1128/aem.68.9.4425-4430.2002

Chen Y, Sible JC, McNabb FMA (2008) Effects of maternal exposure to ammonium perchlorate on thyroid function and the expression of thyroid-responsive genes in Japanese quail embryos. Gen Comp Endocrinol 159:196–207

Chen DP, Yu C, Chang C-Y et al (2012) Branched polymeric media: perchlorate-selective resins from hyperbranched polyethyleneimine. Environ Sci Technol 46:10718–10726. https://doi.org/10.1021/es301418j

Chung J, Shin S, Oh J (2009) Characterization of a microbial community capable of reducing perchlorate and nitrate in high salinity. Biotechnol Lett 31:959–966. https://doi.org/10.1007/s10529-009-9960-1

Coates JD, Achenbach LA (2004) Microbial perchlorate reduction: rocket-fuelled metabolism. Nat Rev Microbiol 2:569–580

Coates JD, Achenbach LA (2006) The microbiology of perchlorate reduction and its bioremediative application. In: Perchlorate. Springer, pp 279–295

Coates J, Jackson WA (2009) Principles of perchlorate treatment. In: Stroo HF, Ward CH (eds) In situ bioremediation of perchlorate in groundwater. Springer New York, pp. 29–53

Coates JD, Michaelidou U, Bruce RA et al (1999) Ubiquity and diversity of dissimilatory (per) chlorate-reducing bacteria. Appl Environ Microbiol 65:5234–5241

Coates JD, Michaelidou U, O'Connor SM et al (2000) The diverse microbiology of (per) chlorate reduction. In: Perchlorate in the environment. Springer, pp 257–270

Dasgupta PK, Martinelango PK, Jackson WA et al (2005) The origin of naturally occurring perchlorate: the role of atmospheric processes. Environ Sci Technol 39:1569–1575

EPA Environmental Protection Agency (2005) EPA sets reference dose for perchlorate

Furdui VI, Zheng J, Furdui A (2018) Anthropogenic perchlorate increases since 1980 in the Canadian high arctic. Environ Sci Technol 52:972–981. https://doi.org/10.1021/acs.est.7b03132

Gan Z, Sun H, Wang R, Deng Y (2014) Occurrence and exposure evaluation of perchlorate in outdoor dust and soil in mainland China. Sci Total Environ 470–471:99–106. https://doi.org/10.1016/j.scitotenv.2013.09.067

Gholamian F, Sheikh-Mohseni MA, Salavati-Niasari M (2011) Highly selective determination of perchlorate by a novel potentiometric sensor based on a synthesized complex of copper. Mater Sci Eng C 31:1688–1691. https://doi.org/10.1016/j.msec.2011.07.017

Ghosh A, Pakshirajan K, Ghosh PK, Sahoo NK (2011) Perchlorate degradation using an indigenous microbial consortium predominantly Burkholderia sp. J Hazard Mater 187:133–139. https://doi.org/10.1016/j.jhazmat.2010.12.130

Ghosh A, Pakshirajan K, Ghosh PK (2014) Bioremediation of perchlorate contaminated environment. In: Singh SN (Ed) Biological remediation of explosive residues. Springer International Publishing, Cham, Switzerland. pp 163–178

Goleman WL, Carr JA, Anderson TA (2002) Environmentally relevant concentrations of ammonium perchlorate inhibit thyroid function and alter sex ratios in developing *Xenopus laevis*. Environ Toxicol Chem 21:590–597

Gu B, Brown GM (2006) Recent advances in ion exchange for perchlorate treatment, recovery and destruction. In: Perchlorate. Springer, pp 209–251

Hatzinger PB (2005) Perchlorate biodegradation for water treatment. Environ Sci Technol 39:239A–247A

He H, Gao H, Chen G et al (2013) Effects of perchlorate on growth of four wetland plants and its accumulation in plant tissues. Environ Sci Pollut Res Int 20:7301–7308. https://doi.org/10.1007/s11356-013-1744-4

Her N, Jeong H, Kim J, Yoon Y (2011) Occurrence of perchlorate in drinking water and seawater in South Korea. Arch Environ Contam Toxicol 61:166–172

Hubé D, Urban S (2013) Préliminaire sur la présence des ions perchlorates dans les eaux souterraines en Alsace

Hurley KD, Shapley JR (2007) Efficient heterogeneous catalytic reduction of perchlorate in water. Environ Sci Technol 41:2044–2049. https://doi.org/10.1021/es0624218

Interstate Technology Regulatory Council (2005) Perchlorate: overview of issues, status, and remedial options

Jackson WA, Böhlke JK, Gu B et al (2010) Isotopic composition and origin of indigenous natural perchlorate and co-occurring nitrate in the southwestern United States. Environ Sci Technol 44:4869–4876

Jackson WA, Davila AF, Estrada N et al (2012) Perchlorate and chlorate biogeochemistry in ice-covered lakes of the McMurdo Dry Valleys, Antarctica. Geochim Cosmochim Acta 98:19–30. https://doi.org/10.1016/j.gca.2012.09.014

Jackson WA, Böhlke JK, Andraski BJ et al (2015) Global patterns and environmental controls of perchlorate and nitrate co-occurrence in arid and semi-arid environments. Geochim Cosmochim Acta 164:502–522. https://doi.org/10.1016/j.gca.2015.05.016

Kounaves SP, Stroble ST, Anderson RM et al (2010) Discovery of natural perchlorate in the Antarctic dry valleys and its global implications. Environ Sci Technol 44:2360–2364. https://doi.org/10.1021/es9033606

Kumarathilaka P, Oze C, Indraratne SP, Vithanage M (2016) Perchlorate as an emerging contaminant in soil, water and food. Chemosphere 150:667–677

Lee J-W, Oh S-H, Oh J-E (2012) Monitoring of perchlorate in diverse foods and its estimated dietary exposure for Korea populations. J Hazard Mater 243:52–58

Lee S, Ji K, Choi K (2014) Effects of water temperature on perchlorate toxicity to the thyroid and reproductive system of *Oryzias latipes*. Ecotoxicol Environ Saf 108:311–317. https://doi.org/10.1016/j.ecoenv.2014.07.016

Li K, Guo J, Li H et al (2019) A combined heterotrophic and sulfur-based autotrophic process to reduce high concentration perchlorate via anaerobic baffled reactors: performance advantages of a step-feeding strategy. Bioresour Technol 279:297–306. https://doi.org/10.1016/j.biortech.2019.01.111

Lindqvist MH, Johansson N, Nilsson T, Rova M (2012) Expression of chlorite dismutase and chlorate reductase in the presence of oxygen and/or chlorate as the terminal electron acceptor in *Ideonella dechloratans*. Appl Environ Microbiol 78:4380–4385. https://doi.org/10.1128/AEM.07303-11

Liu F, Gentles A, Theodorakis CW (2008) Arsenate and perchlorate toxicity, growth effects, and thyroid histopathology in hypothyroid zebrafish *Danio rerio*. Chemosphere 71:1369–1376

Logan BE, Wu J, Unz RF (2001) Biological perchlorate reduction in high-salinity solutions. Water Res 35:3034–3038. https://doi.org/10.1016/S0043-1354(01)00013-6

Maffini MV, Trasande L, Neltner TG (2016) Perchlorate and diet: human exposures, risks, and mitigation strategies. Curr Environ Heal Reports 3:107–117. https://doi.org/10.1007/s40572-016-0090-3

Matsubara T, Fujishima K, Saltikov CW et al (2016) Earth analogues for past and future life on Mars: isolation of perchlorate resistant halophiles from Big Soda Lake. Int J Astrobiol 16:218–228. https://doi.org/10.1017/S1473550416000458

McNabb AFM (2003) The effects of perchlorate on developing and adult birds. SERDP, Strategic Environmental Research and Development Program

Morreale de Escobar G, Jesús Obregón M, Escobar del Rey F (2000) Is neuropsychological development related to maternal hypothyroidism or to maternal hypothyroxinemia? J Clin Endocrinol Metab 85:3975–3987. https://doi.org/10.1210/jcem.85.11.6961

Murray CW, Bolger PM (2014) Environmental contaminants: perchlorate. In: Motarjemi Y (ed) Encyclopedia of food safety. Academic, Waltham, pp 337–341

Na C, Cannon FS, Hagerup B (2002) Perchlorate removal via iron-preloaded GAC and borohydride regeneration. J Am Water Works Ass 94:90–102. https://doi.org/10.1002/j.1551-8833.2002.tb10233.x

Okeke BC, Giblin T, Frankenberger WT (2002) Reduction of perchlorate and nitrate by salt tolerant bacteria. Environ Pollut 118:357–363. https://doi.org/10.1016/S0269-7491(01)00288-3

Parette R, Cannon FS, Weeks K (2005) Removing low ppb level perchlorate, RDX, and HMX from groundwater with cetyltrimethylammonium chloride (CTAC) pre-loaded activated carbon. Water Res 39:4683–4692. https://doi.org/10.1016/j.watres.2005.09.014

Pleus RC, Corey LM (2018) Environmental exposure to perchlorate: a review of toxicology and human health. Toxicol Appl Pharmacol 358:102–109. https://doi.org/10.1016/j.taap.2018.09.001

Rikken GB, Kroon AGM, Van Ginkel CG (1996) Transformation of (per) chlorate into chloride by a newly isolated bacterium: reduction and dismutation. Appl Microbiol Biotechnol 45:420–426

Ryu HW, Nor SJ, Moon KE et al (2012) Reduction of perchlorate by salt tolerant bacterial consortia. Bioresour Technol 103:279–285. https://doi.org/10.1016/j.biortech.2011.09.115

Sanchez CA, Krieger RI, Khandaker NR et al (2006) Potential perchlorate exposure from Citrus sp. irrigated with contaminated water. Anal Chim Acta 567:33–38

Sankar S, Prajeesh GP, Anupama VN et al (2014) Bifunctional lanthanum phosphate substrates as novel adsorbents and biocatalyst supports for perchlorate removal. J Hazard Mater 275:222–229. https://doi.org/10.1016/j.jhazmat.2014.04.046

Schmidt F, Schnurr S, Wolf R, Braunbeck T (2012) Effects of the anti-thyroidal compound potassium-perchlorate on the thyroid system of the zebrafish. Aquat Toxicol 109:47–58. https://doi.org/10.1016/j.aquatox.2011.11.004

Shrout JD, Scheetz TE, Casavant TL, Parkin GF (2005) Isolation and characterization of autotrophic, hydrogen utilizing, perchlorate-reducing bacteria. Appl Microbiol Biotechnol 67:261–268

Simonaitis R, Heicklen J (1975) Perchloric acid: a possible sink for stratospheric chlorine. Planet Space Sci 23:1567–1569

Srinivasan R, Sorial GA (2009) Treatment of perchlorate in drinking water: a critical review. Sep Purif Technol 69:7–21. https://doi.org/10.1016/j.seppur.2009.06.025

Steinmaus CM (2016) Perchlorate in water supplies: sources, exposures, and health effects. Curr Environ Heal Rep 3:136–143. https://doi.org/10.1007/s40572-016-0087-y

Stepanov VG, Xiao Y, Tran Q et al (2014) The presence of nitrate dramatically changed the predominant microbial community in perchlorate degrading cultures under saline conditions. BMC Microbiol 14:225

Sun Y, Gustavson RL, Ali N et al (2009) Behavioral response of dissimilatory perchlorate-reducing bacteria to different electron acceptors. Appl Microbiol Biotechnol 84:955–963. https://doi.org/10.1007/s00253-009-2051-3

Theis TL, Zander AK, Li X et al (2002) Electrochemical and photocatalytic reduction of perchlorate ion. J Water Supply Res Technol 51:367–374. https://doi.org/10.2166/aqua.2002.0033

Thrash JC, Van Trump JI, Weber KA et al (2007) Electrochemical stimulation of microbial perchlorate reduction. Environ Sci Technol 41:1740–1746

Ting W-P, Lu M-C, Huang Y-H (2008) The reactor design and comparison of Fenton, electro-Fenton and photoelectro-Fenton processes for mineralization of benzene sulfonic acid (BSA). J Hazard Mater 156:421–427. https://doi.org/10.1016/j.jhazmat.2007.12.031

U.S. Army Center for Health Promotion and Preventive Medicine, Program HERP& EHRA (2007) Wildlife toxicity assessment for perchlorate. USCHPPM U.S. Army Center for Health Promotion and Preventive Medicine

U.S. Food and Drug Administration (2008) US Food and Drug Administration's Total Diet

Urbansky ET (2002) Perchlorate as an environmental contaminant. Environ Sci Pollut Res 9:187–192. https://doi.org/10.1007/BF02987487

Urbansky ET, Brown SK, Magnuson ML, Kelty CA (2001) Perchlorate levels in samples of sodium nitrate fertilizer derived from Chilean caliche. Environ Pollut 112:299–302

Vellanki BP, Batchelor B, Abdel-Wahab A (2013) Advanced reduction processes: a new class of treatment processes. Environ Eng Sci 30:264–271

Vigliotta G, Motta O, Guarino F et al (2010) Assessment of perchlorate-reducing bacteria in a highly polluted river. Int J Hyg Environ Health 213:437–443. https://doi.org/10.1016/j.ijheh.2010.08.001

Vijaya Nadaraja A, Gangadharan Puthiya Veetil P, Bhaskaran K (2013) Perchlorate reduction by an isolated *Serratia marcescens* strain under high salt and extreme pH. FEMS Microbiol Lett 339:117–121

Volesky B (1999) Biosorption for the next century. Process Metall 9:161–170

Wallace W, Ward T, Breen A, Attaway H (1996) Identification of an anaerobic bacterium which reduces perchlorate and chlorate as *Wolinella succinogenes*. J Ind Microbiol 16:68–72

Waller AS, Cox EE, Edwards EA (2004) Perchlorate-reducing microorganisms isolated from contaminated sites. Environ Microbiol 6:517–527. https://doi.org/10.1111/j.1462-2920.2004.00598.x

Wan Y, Wu Q, Abualnaja KO et al (2015) Occurrence of perchlorate in indoor dust from the United States and eleven other countries: implications for human exposure. Environ Int 75:166–171. https://doi.org/10.1016/j.envint.2014.11.005

Wan D, Liu Y, Niu Z et al (2016) Perchlorate reduction by hydrogen autotrophic bacteria and microbial community analysis using high-throughput sequencing. Biodegradation 27:47–57

Wang C, Lippincott L, Meng X (2008a) Kinetics of biological perchlorate reduction and pH effect. J Hazard Mater 153:663–669. https://doi.org/10.1016/j.jhazmat.2007.09.010

Wang DM, Shah SI, Chen JG, Huang CP (2008b) Catalytic reduction of perchlorate by H2 gas in dilute aqueous solutions. Sep Purif Technol 60:14–21. https://doi.org/10.1016/j.seppur.2007.07.039

Wang Z, Gao M, Zhang Y et al (2014) Perchlorate reduction by hydrogen autotrophic bacteria in a bioelectrochemical reactor. J Environ Manage 142:10–16. https://doi.org/10.1016/j.jenvman.2014.04.003

Wu Q, Zhang T, Sun H, Kannan K (2010) Perchlorate in tap water, groundwater, surface waters, and bottled water from China and its association with other inorganic anions and with disinfection byproducts. Arch Environ Contam Toxicol 58:543–550

Wu Q, Oldi JF, Kannan K (2011) Fate of perchlorate in a man-made reflecting pond following a fireworks display in Albany, New York, USA. Environ Toxicol Chem 30:2449–2455

Xiao Y, Roberts DJ (2013) Kinetics analysis of a salt-tolerant perchlorate-reducing bacterium: effects of sodium, magnesium, and nitrate. Environ Sci Technol 47:8666–8673

Xie Y, Tao G, Chen Q, Tian X (2014) Effects of perchlorate stress on growth and physiological characteristics of rice (*Oryza sativa* L.) seedlings. Water Air Soil Pollut 225:2077. https://doi.org/10.1007/s11270-014-2077-8

Xu J, Logan BE (2003) Measurement of chlorite dismutase activities in perchlorate respiring bacteria. J Microbiol Methods 54:239–247. https://doi.org/10.1016/s0167-7012(03)00058-7

Xu X, Gao B, Jin B et al (2015) Study of microbial perchlorate reduction: considering of multiple pH, electron acceptors and donors. J Hazard Mater 285:228–235. https://doi.org/10.1016/j.jhazmat.2014.10.061

Ye L, You H, Yao J, Su H (2012) Water treatment technologies for perchlorate: a review. Desalination 298:1–12. https://doi.org/10.1016/j.desal.2012.05.006

Ye L, You H, Yao J et al (2013) Seasonal variation and factors influencing perchlorate in water, snow, soil and corns in Northeastern China. Chemosphere 90:2493–2498

Yoon J, Yoon Y, Amy G, Her N (2005) Determination of perchlorate rejection and associated inorganic fouling (scaling) for reverse osmosis and nanofiltration membranes under various operating conditions. J Environ Eng 131:726–733

Youngblut MD, Tsai CL, Clark IC et al (2016) Perchlorate reductase is distinguished by active site aromatic gate residues. J Biol Chem 291:9190–9202. https://doi.org/10.1074/jbc.M116.714618

Zhang T, Wu Q, Sun HW et al (2010) Perchlorate and iodide in whole blood samples from infants, children, and adults in Nanchang, China. Environ Sci Technol 44:6947–6953

Zhang T, Ma Y, Wang D et al (2016) Placental transfer of and infantile exposure to perchlorate. Chemosphere 144:948–954

Zhu Y, Gao N, Chu W et al (2016) Bacterial reduction of highly concentrated perchlorate: kinetics and influence of co-existing electron acceptors, temperature, pH and electron donors. Chemosphere 148:188–194

Sources of Antibiotic Resistant Bacteria (ARB) and Antibiotic Resistance Genes (ARGs) in the Soil: A Review of the Spreading Mechanism and Human Health Risks

Brim Stevy Ondon, Shengnan Li, Qixing Zhou, and Fengxiang Li

Contents

1 Introduction	123
2 ARGs in the Soil Environment	124
3 Sources of ARGs in the Soil Environment	125
3.1 ARB and ARGs in Soils Due to Manure Application	125
3.2 Different Types of Manure Sources Influence ARGs Occurrence in the Soil	128
3.3 Wastewater as a Source of ARB and ARGs in the Soil	129
4 Dissemination of Associated-ARGs with Anthropogenically Impacted Environments into the Soil	135
4.1 Human Exposure Risk Related to Runoff and Leaching	136
4.2 Human Risk Exposure Pathways Due to Plants Uptake	137
4.3 Human Exposure Pathways Due to Groundwater Contaminated with Antibiotics and Their Related ARB and ARGs	138
5 Sorption of Antibiotics in Soil: ARB and ARGs Dissemination	139
5.1 Transfer of Antibiotics Resistance Genes in Agroecosystem	140
5.2 ARB and ARGs Associated Genes in Livestock Production Farms	141
6 Environmental and Ecological Effects of Soil Antibiotics and Related ARB and ARGs	141
7 Health Effects of Soil Contaminated Antibiotics and Their Related ARB and ARGs	142
7.1 Impact of Antibiotic Resistance on Human and Environmental Health	143
7.2 Ecological Health Risk Related to the Antibiotics and the Antibiotic Resistance	144
8 Suggestions and Mitigation Strategies	144
9 Conclusion	146
References	147

Abstract Soil is an essential part of our ecosystem and plays a crucial role as a nutrient source, provides habitat for plants and other organisms. Overuse of antibiotics has accelerated the development and dissemination of antibiotic resistant

---

B. S. Ondon · S. Li · Q. Zhou · F. Li (✉)
Key Laboratory of Pollution Processes and Environmental Criteria at Ministry of Education, Tianjin Key Laboratory of Environmental Remediation and Pollution Control, College of Environmental Science and Engineering, Nankai University, Tianjin, People's Republic of China
e-mail: lifx@nankai.edu.cn

bacteria (ARB) and antibiotic resistance genes (ARGs). ARB and ARGs are recognized as emerging environmental contaminants causing soil pollution and serious risks to public health. ARB and ARGs are discharged into soils through several pathways. Application of manure in agriculture is one of the primary sources of ARB and ARGs dissemination in the soil. Different sources of contamination by ARB and ARGs were reviewed and analyzed as well as dissemination mechanisms in the soil. The effects of ARB and ARGs on soil bacterial community were evaluated. Furthermore, the impact of different sources of manure on soil microbial diversity as well as the effect of antibiotics on the development of ARB and ARGs in soils was analyzed. Human health risk assessments associated with the spreading of ARB and ARGs in soils were investigated. Finally, recommendations and mitigation strategies were proposed.

Graphical Abstract

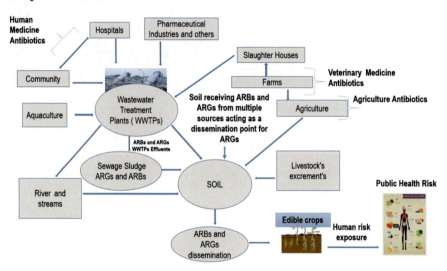

Keywords Agricultural soil · Antibiotic resistance genes · Antibiotic resistant bacteria · Emerging environmental contaminants · Soil environment

Abbreviations

AS	Activated sludge
ARB	Antibiotic resistant bacteria
ARs	Antibiotic resistomes
EECs	Emerging environmental contaminants
HGT	Horizontal gene transfer
LGT	Lateral gene transfer
MGEs	Mobile genetic elements

SS   Sewage sludge
WWTPs   Wastewater treatment plants
AMRs   Antimicrobial resistances
STPs   Sludge treatment plants

Highlights
- Recent studies on occurrence of ARB and ARGs in soils were reviewed.
- Effects of different contamination sources on the diversity and abundance of ARB and ARGs in soils were evaluated.
- The effects of the source and type of manure on soil microbial community and spread of ARGs were analyzed.
- Human health risks due to dissemination of ARB and ARGs in soils were critically reviewed.
- Recommendations, perspectives, and mitigation measures were proposed.

1 Introduction

Antibiotic resistance genes (ARGs) are emerging environmental pollutants (Pruden et al. 2006). ARGs encoding for antibiotic resistance are present in several environments including soils, water, and sediments (Chen and Zhang 2013; Knapp et al. 2010b, 2011). ARB and ARG contaminants from environmental bacteria to human pathogens pose serious risks to the public health (Pehrsson et al. 2016a, b). Certain antibiotics are no longer used due to the high prevalence of resistance (Wenzel and Edmond 2000), and infections caused by bacterial contamination are still claiming hundreds of thousands of lives globally (Nathan et al. 2004). The global consumption of antibiotics has increased considerably over the past few decades. The use of antibiotics largely results in increased production of ARGs, which subsequently poses a great risk to human health globally. Soil is considered as the largest reservoir of ARB and ARGs (Nesme and Simonet 2015; Nesme et al. 2015). A wide variety of novel ARGs are regularly identified in the soils and they exhibit major resistance mechanisms, which have increased our knowledge of antibiotic resistomes (ARs) in the environment (Forsberg et al. 2014a, b). Analysis of archived soils revealed that ARGs abundance substantially increased in soils since the antibiotics era began in the 1940s (Knapp et al. 2010b, 2011).

Antibiotics and associated ARGs can enter soil environments through several ways. Application of manure and irrigation of soil using wastewater effluents are generally the two key ways through which ARB and ARGs are transferred into soils. Wastewater treatment plants (WWTPs) have been identified as potential hotspots that facilitate the spread of ARB and ARGs (Rizzo et al. 2013). Effluents from WWTPs contain a significant amount of ARB and ARGs. This phenomenon has been of particular concern in several developing countries with a compelling need to reuse treated wastewater. Reuse of treated wastewater effluents for agricultural irrigation poses public health risks because of the contamination caused by ARB and ARGs. The practice of applying manure to agricultural soils introduces not only nutrients required to maintain soil fertility but also antibiotics and associated ARB and ARGs into the soil. Antibiotics develop resistance once in the soil and can have adverse impacts on the soil ecosystem (Hashmi et al. 2017). However, soils inherently contain a baseline abundance of ARB and ARGs, which are considered as contaminants of emerging concern due to their high toxicity and persistence in the soil.

This review provides an overview of recent studies and pathways for possible sources of soil contamination by antibiotics as well as associated ARB and ARGs. The present study evaluated the mechanisms and different soil contamination pathways, and reviewed the fate of antibiotics and ARGs in the soil. The effects and human health risks associated with the spread of ARB and ARGs in soils were analyzed, and corresponding mitigation strategies were proposed.

2 ARGs in the Soil Environment

Conventionally, the presence of ARB and ARGs in natural environments is believed to stem from anthropogenic activities such as sewage discharge and animal husbandry coupled with selective pressure from clinically and agriculturally derived antibiotic compounds. However, the realization that numerous ARGs are present in terrestrial and aquatic microbiomes and the fact that the genes predated antibiotic use by millions of years necessitate a more holistic approach to assess the distribution, abundance, and dynamics of ARGs in soil ecosystems (Lebreton et al. 2017). The current knowledge on ARGs in native soils and the impact of anthropogenic activities on ARGs dissemination in soils are analyzed in the subsequent sections.

ARB and ARGs occur in different types of soils globally (D'Costa et al. 2011). Numerous studies have revealed that various types of ARGs significantly increased in soils, with tetracycline resistant genes being one of the key resistant genes exhibiting the highest increase rate (Knapp et al. 2010a). Globally, tetracycline resistant genes that are abundant in the soil occur in the order of 10^{-7} to 10^{-2} genes copies/16S rRNA (Ji et al. 2012; Wang et al. 2015). Some studies have been conducted on sulfonamide (*sul*) resistance genes, showing the presence of the presence of *sul* genes and its relative abundance in soils of certain developing countries (Munir and Xagoraraki 2011). Table 1 presents the literature review summary of soil ARGs and their relative abundances.

Table 1 Different types of soil and the relative abundance of ARGs

Type of soil	ARGs	Relative abundance	Reference
Ditched irrigated and mineral phosphate fertilized soil	*tetM, tetW, tetQ, tetO*	10^{-4}–10^{-2}	Knapp et al. (2010a)
Unamended and sewage sludge-amended soil	*tetM, tetQ, tetW*	10^{-5}–10^{-2}	Knapp et al. (2011)
Manure and biosolids fertilized soil	*sul1*	10^{-6}–10^{-5}	Munir and Xagoraraki (2011)
	tetW, tetO	10^{-7}–10^{-4}	
Soil in residential areas	*Sul1, sul2*	10^{-8}–10^{-2}	Knapp et al. (2017)
	TetM, tetW	10^{-9}–10^{-2}	
Soil adjacent to swine feedlots	*tetM, tetO*	10^{-5}–10^{-2}	Wu et al. (2010)
	tetQ, tetW		
Soil adjacent to livestock feedlots	*sul1, sul2*	10^{-5}–10^{-2}	Ji et al. (2012)
	tetB/P, tetO, tetM, tetW,	10^{-8}–10^{-2}	
Soil in livestock feedlots	*sul1, sul2*	10^{-4}–10^{-3}	Wang et al. (2015)
	tetM, tetO, tetW, tetB/P	10^{-5}–10^{-3}	
Agricultural soil	*sul1, sul2*	10^{-6}–10^{-2}	Zhou et al. (2017)
	tetM, tetW, tetQ, tetO, tetT, tetB/P	10^{-8}–10^{-2}	

3 Sources of ARGs in the Soil Environment

As mentioned in the previous sections, ARB and ARGs are transferred into the soil through several pathways. Table 2 summarizes the different sources of soil contamination by ARGs and ARB.

3.1 *ARB and ARGs in Soils Due to Manure Application*

Application of manure is a principal source of ARB and ARGs in the soil, and leads to the accumulation of large quantities of ARB and ARGs in the surrounding soil environment (Yang et al. 2014; Ding et al. 2014). This practice is commonly used because it provides nutrients that are necessary for soil fertilization. On the other hand, manures are known to be the hotspots of ARB and ARGs (Oliver et al. 2020; Wallace et al. 2018). Qian et al. also revealed the presence of almost 109 ARGs in manure and compost from chicken, bovine, and pig farms (Qian et al. 2018). Animal manures are composed of animal excrements containing high levels of ARB and ARGs. The overuse of antibiotics in animal husbandry is the main reason of the increase of ARB and ARGs in animal excrements. The practice of soil fertilization with animal manures is unfortunately the main reason for the dissemination of ARB and ARGs in soils. Therefore, monitoring of AR levels should begin at the manure

Table 2 Different sources of antibiotic resistance genes and bacteria in the soil

Source	Types	Contamination	ARGs or ARB impact on receiving soil and	Reference
Livestock's excrements	Animal manure	Direct application of manures in soil	Tetracycline resistance conferred by efflux pumps (*tetA, tetB, tetC, tetG, tetH, tetL, tetZ*) and those conferred by ribosomal protection proteins (*tetM, tetO, tetQ, tetW*) were present in high abundances per mass or volume of animal manure	Wang et al. (2015)
Land application of animal manures	Soil microcosm	Direct land application of manures	Considerably high *sul1, sul2, tetB, tetM, ermB, ermF, fexA, cfr*	Wang et al. (2018)
	Chicken manure	Farmland soil chronically fertilized with chicken manure	High level of many types of resistant bacteria and ten ARGs (*tetW, tetO, tetT, tetM, tetA, tetL, tetQ, sul1, sul2* and *sul3*)	Zhao et al. (2017)
	Swine and dairy manures	Cornfield and pasture soils receiving swine and dairy manures	89 ARGs were found in the soil. The grassland manured soils had greater ARGs diversity. ARGs such as *intI1-1, vanC, and pncA* showed higher relative quantity in the soil	Chen et al. (2019b)
	Pig manure	Application of pig manure in soil	Occurrence of 38 ARGs and abundance of seven ARGs (*tetL, tetB (P), tetO, tetW, sul1, ermB, and ermF*) significantly increased by long-term exposure of pig manure	Peng et al. (2017)
	Manure-derived amendments	Soils receiving manure-derived amendments	Increased the diversity (richness) of ARGs in soils ($p < 0.01$) and resulted in distinct abundances of individual ARG types. Antibiotic-manure significantly increased the total ARGs relative abundances in the soils. 282 ARGs reduced 4.33-fold up to 307-fold while 210 ARGs increased 2.89-fold up to 76-fold in the antibiotic manure-amended soils	Chen et al. (2019a)
Hospitals hospital wastewater		Irrigation with WWTPs effluents	*VanA, blaIMP; blaOXA; intI1, blaVIM; intI1, blaVIM*	Vaz-Moreira et al. (2016)

Wastewater	Dairy wastewater	Soil irrigated with frequently dairy wastewater	blaCTX-M-1, erm (B), sulI, tet (B), tet (M), and tet (X) and a class 1 integron-integrase gene. Only sulI and tet(X) were detected in soil (3 out of 32 samples) before the wastewater treatments were applied. The occurrence and relative abundance of most genes erm(B), intI1, sulI, and tet(M) increased dramatically	Dungan et al. (2018)
	Municipal wastes	Application of sewage sludge to agricultural soil	ARGs and MGEs were found in soils receiving municipal waste tnpA-07, aac(6'), tet(32), tetPA, tetT, vanB-01, tnpA-07	Urra et al. (2019)
	Treated municipal wastewater effluent	Soil irrigated frequently with municipal wastewater effluent	The concentrations of antibiotics and abundances of resistance genes were significantly greater in irrigated soils, indicating that agricultural activities enhanced the occurrence of antibiotics and resistance genes in the soils tetA/16S, tetB/16S, tet C/16S, tetD/16S, tetE/16S, tetG/16S, tetK/16S, tetL/16S, tetM/16S, tetO/16S, tetS/16S, tetQ/16S, tetX/16S, sulI/16S, sulII/16S, sulIII/16S	Chen et al. (2014)
		Soil irrigated	Occurrence and distribution of tetracycline (TC) and sulfamethazine (SMZ), and the corresponding antibiotic resistance genes (ARGs in six agricultural sites in the Pearl River	
Manure-borne	Indigenous soil		Manure application significantly increased the diversity and abundance of ARGs in soil. Manure-borne microorganisms contributed largely to the elevation of ARGs. In contrast, indigenous soil microorganisms prevent the dissemination of ARGs from manure to soil	Chen et al. (2017)
Sewage sludge	Agricultural soil	Application of sewage sludge to agricultural soil	Soil microbial quality as reflected by the value of the soil quality index was higher in sewage sludge-amended soils. In contrast, the composition of soil prokaryotic communities was not significantly altered	Urra et al. (2019)

level, and a critical control point for the isolation and remediation of antibiotic resistances (ARs) before ARB and ARGs are transported and deposited in agroecosystems and agricultural soils (Pruden et al. 2013).

3.2 Different Types of Manure Sources Influence ARGs Occurrence in the Soil

Manures that are applied in agricultural soils are from different origins. There are three types of manures, including farmyard manure (animal manure), green manure (from plants), and compost manure (decomposed waste materials). The effect of manure application on soil may differ due to their origin and nature. Animal manures are generally the most popular and used in agriculture compared to other types of manures (Chen et al. 2019b). Animal manures used in agricultural fields contain ARB and ARGs that can affect ARs in agricultural soils. Application of particular antibiotics can facilitate bacteriophage mediated transfer of ARB and ARGs in agricultural soil microbiomes. Typical antibiotics that are widely used in livestock production are tetracycline, chlortetracycline, oxytetracycline, and sulfonamides. These antibiotics can be found in animal excrements that are used as manure in agriculture soils. Tetracycline resistance genes (*tetA, tetG, tetM, tetO, tetQ, tetW*, and others) have been detected in soil samples (Wu et al. 2010). The fate of manure-associated antimicrobials applied to the soil can be affected by the sorption properties of soil particles and susceptibility to biotic and abiotic degradation. Table 3 shows the effects of different types of animal manures and the presence of antibiotics, ARB and ARGs in the soil.

Previous research has demonstrated that ARB and ARGs occur abundantly in manured soils. Recent analyses have revealed that ARGs profiles are strongly correlated with the bacterial community composition. These observations indicate that there is a significant correlation between the abundance of MGEs and ARGs, suggesting the existence of a potential effect of horizontal gene transfer on the persistence of ARGs, which should not be overlooked. In addition, soil properties

Table 3 Effects of different types of animal manures and the presence of antibiotics and ARGs in the soil

Types of applied manures	Type and abundance of ARGs present in soil	Reference
Veterinary manure applied	15 tetracycline resistance (*tet*) genes were commonly detected in soil samples, including seven efflux pump genes (*tet*A, *tet*C, *tet*E, *tet*G, *tet*K, *tet*L) and genes (*tet*M, *tet*O, *tet*Q, *tet*S, *tet*T, *tet*W, *tet*B (*tet*X)	Wu et al. (2010)
Manure from chicken pig and bovine	The widespread co-occurrence of ARGs and MGEs. *floR, ermF* were identified	Qian et al. (2018)
Swine manure	Macrolides fluoroquinolones Tetracyclines	Wen et al. (2019)

altered by application of manure could subsequently affect the distribution patterns of ARGs. These findings demonstrate that the temporal patterns and dissemination risk of ARGs in manured soils could contribute to the development of effective strategies to minimize the spread of ARGs in agricultural soils.

3.3 Wastewater as a Source of ARB and ARGs in the Soil

The application of wastewater on agricultural soils is the most widespread method used to irrigate and fertilize the soil. Studies have demonstrated that wastewaters are sources of ARB and ARGs, and they facilitate dissemination of ARB and ARGs into the environment (McKinney et al. 2010; Munir et al. 2011). In addition to the applied nutrients, the contaminants or residues in wastewater accumulate in the soil during irrigation activities (Chee-Sanford et al. 2009a; Chee-Sanford et al. 2009b). Table 4 summarizes the different types of wastewater used for irrigation and their impacts on the soil.

Irrigation of agricultural soils using wastewater effluents results in continuous release of ARB and ARGs into natural and agricultural soil environments (Fatta-Kassinos et al. 2011a, b; Gatica and Cytryn 2013; Negreanu et al. 2012; Rizzo et al. 2013).

3.3.1 Release of ARB and ARGs from WWTP Effluents into the Soil

WWTPs provide conducive environments for the proliferation of ARB and ARGs because of the high concentrations of bacteria in the biofilms (LaPara et al. 2011; Novo and Manaia 2010). Researchers have focused on WWTPs as reservoirs of ARB and ARGs over the last few years. High concentrations of tetracycline *(tetA, tetQ, and tetX)*, sulfonamide *(sulI)* resistance genes, and high levels of ARB have been detected in the final effluents of WWTPs (Liu et al. 2019; Sabri et al. 2020). Moreover, a previous study reported significant levels of ARGs *(tetO, tetW, and sulI)* and ARB in raw sewage, effluents, and biosolid samples of WWTPs (Munir et al. 2011), which were continuously released into the soil. Other crucial ARGs such as *ampC* and *vanA* that are associated with resistance to ampicillin and vancomycin, respectively, have been identified in wastewater samples collected from five municipal WWTPs. Furthermore, several studies have identified the methicillin-resistant gene *mecA* in samples collected from clinical wastewater effluents.

The ARGs present in the final effluents of WWTPs have also been detected in the soil. Figure 1 is an illustration of WWTP effluents containing ARB and ARGs, and the pathways of release into the soil environment.

Table 4 Different wastewater types used for irrigating agricultural soils

Wastewater type	Abundance and ARGs type in wastewater and irrigated soil	Reference
Industrial wastewater	Sulfonamide resistance genes (*sulI* and *sulII*), tetracycline resistance genes (*tetO, tetW* and multiple ARGS including quinolone, sulfonamide, and tetracycline resistance genes, multiple ARGs quinolone, sulfonamide, and tetracycline resistance genes were detected	Li et al. (2017), Lucas et al. (2016) and Yang et al. (2014)
Domestic wastewater	The average abundance of *tet* genes soils ranged from $2.32 \cdot 10^{-6}$ to $1.31 \cdot 10^{-2}$ ARGs copies/16S rRNA genes $0.1.24 \cdot 10^{-6}$ to $9.30 \cdot 10^{-3}$ ARG copies/16S rRNA genes. *tet A, tet C, tet O* in deep soil 3 were detected. *sul* genes with high abundance were also detected. Then five *tet* genes (*tet B, tet E, tet M, tet S,* and *tet X*) which is consistent with their abundances in the irrigation wastewater. *TetA* and *tetO* were the most prevalent tet genes with the highest concentration in irrigation wastewater and irrigated soil samples	Chen et al. (2014), Pan and Chu (2018) and Pepper et al. (2018)
Dairy wastewater	Occurrence and abundance of antibiotic resistance genes *blaCTX-M-1, erm(B), sul1, tet(B), tet (M),* and *tet(X)*. *Sul1* and *tet(X)* were detected in soil before the wastewater treatments were applied. Occurrence and relative abundance of 16S rRNA gene copies. The prevalence of ARGs with high levels of genetic mobile elements in the dairy farms	Cao et al. (2013), Chen et al. (2019b) and Wallace et al. (2018)
Swine wastewater	ARGs (*sul2, sul1, tetG,* and *ermF*) with very high concentrations of copies were detected. The ARGs abundance were lower in winter than in summer time except *ermB* and *mefA*. The gene copies of *tetG, ermF, sul1, sul2, tetX,* and *tetM* were decreased considerably	Sui et al. (2016) and Yang et al. (2014)
Sewage sludge	*mefA, mexB, mexD, mexF, ermF, tetM tetO, tetQ, tetM tetO, tetQ*	Yang et al. (2014) and Zhang et al. (2015)

3.3.2 Impacts of Irrigation Using WWTP Effluents on the Abundance of ARB and ARGs in the Soil

The impact of irrigation using WWTP effluents with high concentrations of ARB and ARGs on the soil has been the primary focus of several researchers. For example, in occasions where soil was irrigated with diluted organic media amended with clinical concentrations of antibiotics, no changes were observed in microbial activities and there were minimal changes in the composition of microbial communities compared to the non-antibiotic-amended soils (Gatica and Cytryn 2013). Furthermore, a previous study revealed that environmentally relevant antibiotic

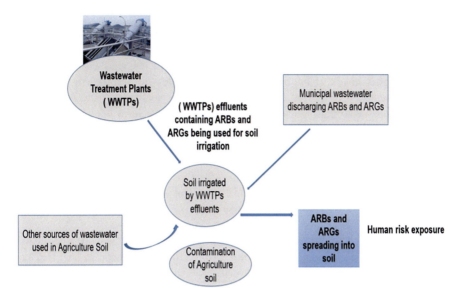

Fig. 1 Release of ARB and ARGs from WWTP effluents into the soil

concentrations do not seemingly exert significant selective pressure on soil microbiomes (Gatica and Cytryn 2013).

Negreanu et al. conducted one of the most significant studies that enhanced our understanding of the effects of irrigation using WWTP effluents on the quantities of soil ARB and ARGs (Negreanu et al. 2012). The study investigated four types of soils with different properties. Irrigation was performed using freshwater or effluents from WWTPs. Resistance to four clinically relevant antibiotics such as tetracycline, erythromycin, sulfonamide, and ciprofloxacin was evaluated. The authors monitored six different types of ARGs that had been previously detected in WWTP effluents. Resistance to several ARGs such as *sul1*, *sul2*, *ermB* and *ermF* genes was analyzed, and results revealed that the relative abundance of resistant isolates and the levels of ARGs were either identical or often higher in freshwaters adjacent to soils irrigated with WWTP effluents (Negreanu et al. 2012). These findings indicated that the concentrations of residual antibiotics associated with WWTP effluents do not exert significant selective pressure that can induce propagation of ARGs.

According to the previously mentioned studies, the large numbers of resistant bacteria entering the soil through WWTP effluents are likely to compete with other bacteria or survive in the soil environment. The high levels of ARB and ARGs observed in both freshwater and soils irrigated with WWTP effluents are predominantly associated with resistomes in native soils. A study by McLain and Williams on the comparison of AR profiles of *Enterococcus* isolated from soils irrigated with WWTP effluents over a long period of time presented similar results (McLain and Williams 2010). The study revealed high levels of ARs in isolates from soils irrigated with both WWTP effluents and freshwater. Although resistance patterns of ARs in the two types of soils differed, isolates from soils irrigated with WWTP effluents exhibited high resistance to lincomycin. Moreover, isolates from a pond filled with groundwater were highly resistant to erythromycin, tetracycline, ciprofloxacin, and tylosin tartrate. Strikingly, isolates from soils irrigated with freshwater exhibited higher levels of multi-resistance than isolates from soils irrigated with WWTP effluents. Nevertheless, the potential effect of selective pressure due to low levels of exposure to antibiotics from unknown sources cannot be excluded. The results indicate a natural occurrence of ARs in soils prior to irrigation using WWTP effluents.

3.3.3 Persistence of ARGs in WWTP Effluents as a Source of ARB and ARGs Dissemination in the Soil

Most methods used to treat effluents from WWTPs are inefficient in relation to complete removal of ARB and ARGs from WWTP effluents (Wang et al. 2017). The effluents from WWTPs containing untreated ARB and ARGs are used for agricultural irrigation, a phenomenon that leads to contamination of agricultural soils. The application of wastewater effluents to agricultural soils is a widely used practice to irrigate crops and recycle nutrients. A recent study conducted by Dungan et al. revealed the presence of several ARGs in soils irrigated with dairy wastewater (Dungan et al. 2018). The authors concluded that the occurrence and abundance of most ARGs (*blaCTX-M-1, erm B, sul1, tetB, tet), tetX*) and a class 1 integron-integrase (*intI1*) in receiving soil increased dramatically after wastewater irrigation. The effluent of wastewater used for soil irrigation can significantly enlarge the reservoir of ARGs in soils. Table 5 summarizes the persistence of ARGs in effluents of WWTPs after treatment.

3.3.4 Application of Sewage Sludge to Agricultural Soils

The application of sewage sludge (SS) as soil amendment is a common agricultural practice. However, sewage sludge from WWTPs and soils amended with sewage sludge have been reported as hotspots for the dissemination of antibiotic resistance. This phenomenon is driven by several factors such as selection pressure exerted by co-exposure to antibiotics and heavy metals. The use of SS as fertilizer is aimed at supplying valuable nutrients (nitrogen-N and phosphorus-P) and organic matter to agricultural soils. Application of SS to agricultural soils enhances the physicochemical and biological properties of soils, while providing plants with essential nutrients (Latare et al. 2014; Lloret et al. 2016; Singh et al. 2012). Soil microbial parameters provide useful information on the activity, biomass, and diversity of soil microbial communities. Nevertheless, the application of SS to agricultural soils poses a great risk to the environment because sewage sludge contains a wide variety of different toxic contaminants (Petrie et al. 2015; Roig et al. 2012). In addition, little consideration has been given to other contaminants present in sewage sludge.

Table 5 Persistence of ARGs in effluents of WWTPs after treatment

Antibiotic resistance genes in WWTP	Persistence of ARGs in effluents after treatment	Reference
Sulfonamide resistance genes (*sulI* and *sulII*), tetracycline resistance genes (*tetO, tetW* and multiple ARGs) including quinolone, sulfonamide, and tetracycline resistance genes	Different ARGs played a critical role in ARG loss during coagulation, presence of different ARGs after treatment	Li et al. (2017)
Multiple ARGs (total = 18) including quinolone, sulfonamide, and tetracycline resistance genes	Multiple ARGs (total = 18) including quinolone, sulfonamide, and tetracycline resistance genes	Yang et al. (2014)
intI1, sul1, tetX, tetG, and 16S rRNA genes	Sulfonamide resistance genes with 98.09% were persistent	Zhang et al. (2016)
ermB, vanA, blaVIM 2 orders of magnitude reduction in *ermB*	Simultaneous increase in *vanA, blaVIM* Alexander	Guo et al. (2017)
ARGs, such as *ermB, tetW, blaTEM, sul I,* and *qnrS* (reduced susceptibility to fluoroquinolones)	The *blaSHV* gene increased in both bioreactors almost to the same extent, in agreement with the assumption that ARGs increase is favored by the presence of selective agents, such as antibiotics	Lucas et al. (2016)
Integrases and genes for resistance to quinolones, chloramphenicol, tetracycline, aminoglycosides, sulfonamides, blactams, and carbapenems	Increase in concentration for resistance most. ARGs were prevalent in all of the samples examined. Ineffective at removing antimicrobial resistance determinants from wastewater (WW)	Wang et al. (2017)

However, contaminants that have been given little consideration in the past and emerging contaminants are becoming a matter of concern. The presence of pharmaceutically active compounds in sewage sludge is particularly a topic that has recently attracted the interest of researchers (Martín et al. 2015). The pharmaceuticals of interest have been antibiotics, due to their considerable medical relevance globally and poor metabolism in both human and animal bodies (Looft et al. 2012; Michael et al. 2013). Antibiotics are not eliminated during wastewater treatment, therefore, they can be deposited in agricultural soils when sewage sludge is applied as fertilizer. Antibiotic degradation in the soil is principally driven by enzymatic transformations conducted by microorganisms (McGrath et al. 1998).

3.3.5 Deposition of ARGs from Livestock Excrements into the Soil Environment

Livestock farming is one of the primary activities through which ARB and ARGs are discharged into the soil. This phenomenon is facilitated by animal excrements that are commonly used as manure in agricultural soils. For example, approximately

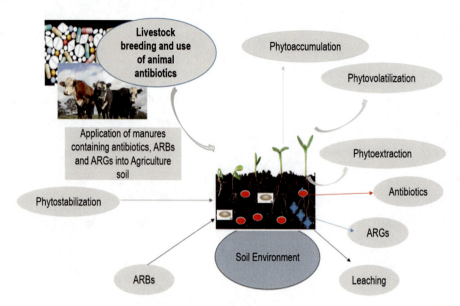

Fig. 2 Pathways through which ARB and ARGs are released from livestock excrement into the soil environment and assimilation by plants

210,000 tons of antibiotics are generated annually in few developing countries. A significant correlation between antibiotic concentration and ARG emergence in feedlot lagoons and soils adjacent to swine farms has been reported (Smith et al. 2004; Wu et al. 2010). ARGs emanate from and multiply in the gastrointestinal tracts of animals. The ARB and ARGs are subsequently excreted by animals through feces, indicating the importance of animal excrements as reservoirs of ARB and ARGs, which are consequently released into the soil environment (Fig. 2).

ARGs such as tetracycline (*tet*) and sulfonamide (*sul*) are frequently detected because of the widespread use of corresponding antibiotics and their persistence in the environment (Gao et al. 2012; Luo et al. 2010). *Tet* and *sul* genes are the most frequently detected ARGs in animal manure and livestock lagoons (Zhu et al. 2013). Numerous studies have revealed that *tet* and *sul* genes are present in soils with relatively high abundance. A study by Chee et al. reported the occurrence and detection of eight genes (*tetO, tetQ, tetW, tetM, tetB, tetS, tetT, and otrA*) in two swine lagoons and subsurface soils (Chee-Sanford et al. 2001), implying that ARGs are discharged into the soil environment through livestock excrements.

3.3.6 ARGs in Native Soils

A study by D'Costa et al. indicated the potential occurrence of soil-borne ARGs. Soil environments contain several types of microbes. Different types of resistomes occur in the soil due to the diverse nature of the microbial community. Native resistance is attributed to the presence of antibiotics that are directly discharged into the soil and the selective pressure of antibiotics on the composition of soil microbial community. A comprehensive study by D'Costa et al. was the first study to reveal the vast dimensions of antibiotic resistance in the soil. Results of the study demonstrated that apparent resistance is associated with new soil-borne ARGs and clinically characterized ARGs. These findings suggested that the capacity of antibiotic resistance in soils is due to the complexity of microbial communities and abundance of antibiotic producing bacteria in the soil. The authors concluded that the bacteria identified as "resistomes" were associated with known and novel ARGs in specific environments. This indicates that resistomes in the soil can be a crucial source of novel ARGs.

In 2011 D'Costa conducted a study to analyze ancient DNA from soils. Ancient permafrost soils strongly supported the resistome hypothesis by demonstrating that soil-associated ARGs predated clinical use of antibiotics. A very diverse collection of ARGs encoding resistance for β-lactam, *tet*, and glycopeptide antibiotics have been detected in soils. These ARGs are very similar to modern pathogen-associated ARGs (D'Costa et al. 2011). Studies have demonstrated that mobilization of ARGs from environmental microbiomes to clinical pathogens can significantly increase ARGs activity under specific conditions.

4 Dissemination of Associated-ARGs with Anthropogenically Impacted Environments into the Soil

Comprehensive studies have been conducted to elucidate characteristics of soil microbiomes as reservoirs for a diverse array of novel and clinically characterized ARGs. Nonetheless, the transient and constant influx of anthropogenic factors in soil ecosystems is believed to be caused by proliferation of soil resistomes (Perry and Wright 2013; Perry et al. 2014). This phenomenon has been associated with dissemination and persistence of ARGs from anthropogenic sources such as wastewater, animal manure, and municipal biosolids. Moreover, residual concentrations of selective elements such as antibiotics, heavy metals, and detergents are linked to anthropogenic activities. ARGs derived from anthropogenic activities can be horizontally transferred to native soil communities or alternatively selective elements can stimulate resistance in native soil resistomes. Although several studies supporting the two concepts have been published (Berglund 2015; Chen et al. 2013; Storteboom et al. 2010; Wang et al. 2016), available data are still inconclusive and do not provide direct evidence that links specific factors to individual ARGs. Other studies have revealed that factors previously unconsidered could also be significant. Overall, the dynamics of soil resistomes are more complex than originally thought. The subsequent section explores how anthropogenic activities affect the diversity and abundance of ARGs in soils.

Soil archives provide a unique opportunity to evaluate the diversity and abundance of ARGs in the soil. A groundbreaking study by Knapp et al. assessed the relative abundances of clinically relevant ARGs in an archived soil collection and they established that the relative abundances of all of the targeted ARGs families significantly increased in the soil since when the antibiotic compounds were first discovered and used (Knapp et al. 2010b). This observation was especially evident for specific tet resistance genes, whose abundance increased in the soil. The study strongly suggested that anthropogenic activities associated with antibiotic use could be responsible for the increase in ARG abundance. However, the present study did not characterize specific factors associated with the observed increase in antibiotic resistance.

4.1 Human Exposure Risk Related to Runoff and Leaching

Transport of antibiotics and their related ARB and ARGs into groundwater or surface water may occur through leaching or runoff. Surface transport of antibiotics with ARB and ARGs via runoff is attributed to delayed infiltration of water into the soil because of surface sealing through manure and particle bound transport (Burkhardt et al. 2005; Kreuzig and Höltge 2005). In addition, antibiotics associated with land application of sludge and livestock manure can also access aquatic environments via surface runoff. These antibiotic residues can then influence the bacterial diversity and microbial dynamics of the aquatic ecosystem. Besides inherent resistance, the occurrence of ARB and ARGs could be influenced by antibiotic residues which exert selective pressure on the microbial community in the environment causing direct risks of contamination to human exposed aquatic environment. Humans are directly exposed to the contaminated water via several activities including direct consumption and contact during aquatic environment activities. Direct exchange between humans and animals as well as environment and humans can cause direct risks of contamination. Therefore, runoff constitutes a major pathway for surface water and aquatic environment contamination.

In a study involving tylosin, the antibiotic was not noticed in soil or leachates after dissolving large amounts of tylosin in slurry (Kay et al. 2005). The mobility rate of antibiotics in the soil depends on several factors including chemical properties, temperature, moisture content, and the timing of manure application. Furthermore, surface application of animal manure can considerably increase the rate of runoff water from the treated field due to the surface sealing effect of manure particles. Rapid movement of antibiotics largely depends on the size of soil macropores where smaller macropores have less significant effect on antibiotics leaching.

Currently, urban runoff water receives less attention than other water bodies with regard to ARB and ARGs. However, results from the few available studies reveal an undeniable role of runoff in the dissemination of antibiotic resistance genes in urban areas (Almakki et al. 2019; Garner et al. 2017; Lee et al. 2020). Soil bacteria are major contributors to aquatic bacterial contamination because they are organized in rich and diverse communities. Since most bacteria are clustered in the top soil layer, they are in a better position to disseminate into water bodies during flood events. On the other hand, soil can be considered as a natural reservoir for ARB and ARGs. Soil is also directly involved in further contamination of groundwater through infiltration and contamination of adjacent surface water bodies through surface runoff. Reports have indicated that the erosion control practice has the ability to control leaching and runoff of antibiotics because it has extremely low aqueous concentration with low losses. This method has been proven to be beneficial in reducing the transport of antibiotics in soil (Davis et al. 2006). Moreover, sulfonamides are water soluble (Hu et al. 2010), thus they can be transferred easily through leaching and runoff leading to high chances of direct contact with human.

4.2 Human Risk Exposure Pathways Due to Plants Uptake

Continuous discharge of antibiotics in soil through repetitive manure application use may ultimately build up high concentrations of ARB and ARGs enough to enter into the terrestrial environment as potent hazards (Bassil et al. 2013). Depending on the concentration and exposure time, different plant organs and tissues respond differently towards antibiotics. Roots, cotyledons, and cotyledon petioles exhibit a toxic effect when exposed to low antibiotic concentrations and toxicity while other parts like internodes and leaf length show an increased growth. Recently, research has been done on the phytoremediation potential of plants against different antibiotics.

Naturally, plants play the role of phytoremediation of toxic materials (like heavy metals and PAHs) from the soil. In recent years, phytoremediation through phytostabilization, phytoextraction, phytovolatilization, and phytoaccumulation is emerging as a new technology which is considered as effective in elimination of antibiotics from the soil (Michelini et al. 2014). Vegetables like corn, potatoes, and lettuce have the ability to absorb antibiotics at different rates when grown in soil fertilized with animal manure. The results are the same with vegetables grown on soil improved with liquid manure containing antibiotics, mainly sulfamethazine, in a greenhouse. Alimentation chain production seems to be the main exposure pathway due to uptake of antibiotics and ARB by plants. Surface water, groundwater, and agricultural soil have become reservoirs for antibiotics due to the current manure managing practices (Hu et al. 2010). Figure 3 shows a system approach that describes human exposure pathways for antibiotics and their related ARB and ARGs from contaminated soil.

Humans risk exposure from antibiotics and ARB transported within soils to ground and surface water by both leaching and runoff. Urban runoffs have significance in environmental antibiotic resistance because the runoff is able to carry bacteria from different environmental media including antimicrobial resistance genes to surface rivers and groundwater. This poses high risks of contamination to humans from direct exposure through bathing and recreational use of water, washing foods and clothes,

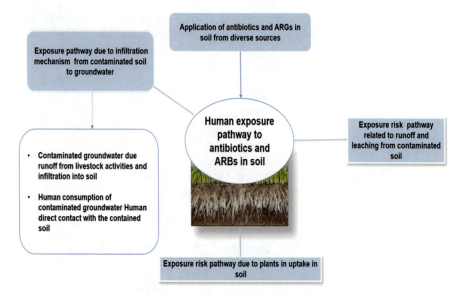

Fig. 3 Approach to human exposure pathways for antibiotics and related ARB and ARGs

contaminated water, and sediments. Plant uptake of antibiotics and antibiotics resistance bacteria drives ARB and ARGs into vegetables that are grown on soil improved with liquid manure containing antibiotics and ARB. Human can also experience direct contact with the contaminated soil by working in agriculture field and consumption of edible root crops (potatoes, carrots, and radishes). These contamination pathways are the main sources of ARB and ARGs contamination in human.

4.3 Human Exposure Pathways Due to Groundwater Contaminated with Antibiotics and Their Related ARB and ARGs

Contamination of groundwater by antibiotics and their related ARB and ARGs is due to runoff from livestock activities, infiltration and leaching into soil. In a study, researchers compared samples from an advanced groundwater treatment facility and groundwater aquifers to detect differences in ARGs concentrations. Results indicated that the groundwater samples had a ubiquitous presence of ARGs in both control locations. New and emerging contaminants like antibiotics and their related ARB and ARGs pose potential hazards to public safety and groundwater water security. Contaminated groundwater may poses serious human health risks related with indirect reuse of contaminated groundwater in our drinking water system being the main common pathway for human exposure. Moreover, groundwater can serve as a source of raw drinking water thereby causing human contamination with ARB and ARGs. Contaminated groundwater can also enter lakes and rivers through infiltration. Construction and operation of extraction wells provides water for agricultural, municipal, and industrial use. This constitutes a potential human exposure risk for populations in permanent contact with agricultural products, municipal, and industrial activities obtained from contaminated groundwater sources.

5 Sorption of Antibiotics in Soil: ARB and ARGs Dissemination

Antibiotics interact with soil organic matter resulting in sorption, binding, attachment, and fixation of the chemicals on the soil matrix. Interaction strength depends on parameters, such as chemical nature of the species, soil characteristics, influence of temperature, humidity, and the soil solution constituents (Kumar et al. 2005). Further, sorption rate of antibiotics on different surfaces is affected by contact time whereas sorption of sulfathiazole and sulfamethazine is affected by high pH levels (Kurwadkar et al. 2007). Sorption of these antibiotics by-products on charged surfaces of mineral and organic exchange sites in soil is driven by electrostatic forces (Holten Lützhøft et al. 2000). Antibiotic related toxicity includes damage of vital microbes that supply nutrients to plants and change of microbial population through resistant selection by altering soil microbial composition (Knapp et al. 2011) (Fig. 4).

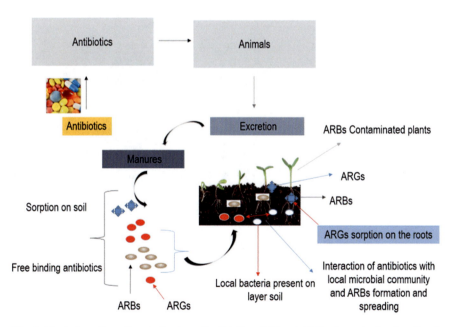

Fig. 4 Illustration of sorption mechanism of antibiotics, ARB and ARGs dissemination in soil and interaction with soil microbial community

5.1 Transfer of Antibiotics Resistance Genes in Agroecosystem

Antibiotics are absorbed into soil through use of antibiotic-contaminated water for irrigation or use of animal manure as fertilizer. Continuous exposure of antibiotics to bacteria found at the upper layer of soil results in generation of ARB and ARGs. ARB and ARGs are eliminated through fecal material and get absorbed into soil from manure during agriculture practice. Moreover, ARB can be taken up by crops or vegetables and are eventually ingested by human during food uptake (Fig. 5). They can also be transferred from animals to human via dairy or meat products. Previous studies have reported that direct contact with cattle can also lead to the spread of ARB from animals to humans (Graham et al. 2019).

Despite advances in antibiotic detection methods, limited data are available on the fate and occurrence of antibiotics, as well as their temporal and spatial distribution in agroecosystems. A predictive model for estimation of expected antibiotic concentration at the landscape-scale has been proposed as an alternative to large-scale and high-cost monitoring programs (Boxall et al. 2003a, b). These models are based on accurate antibiotics usage data as well as mechanistic knowledge of metabolism, transport, fate, landscape, and hydrologic processes. However, usage data is not universally available.

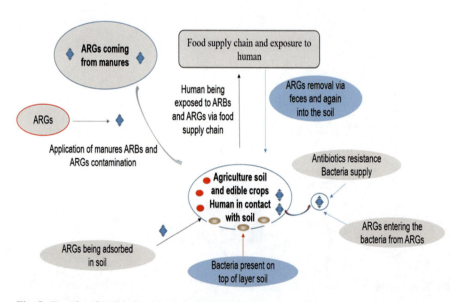

Fig. 5 Transfer of antibiotic resistance genes in agroecosystem

Fig. 6 Occurrence of antibiotic resistance genes in agricultural soils – Gene Exchange Platform (Markus Woegerbauer ICCE 2017)

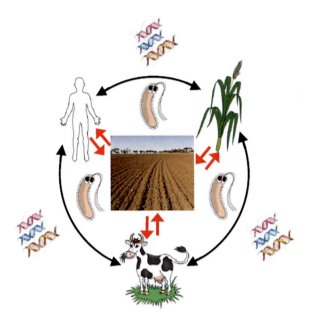

5.2 ARB and ARGs Associated Genes in Livestock Production Farms

Previous studies reported the presence of resistance genes, mainly genes conferring resistance to tetracycline and other antibiotics in both waters and manure-applied soil that were used for livestock production (Graham et al. 2019; Koike et al. 2007). However, chloramphenicol-resistance genes were not detected after stabilization. Notably, groundwater adjacent to the manure pit was susceptible to contamination from the waste lagoon. Studies monitoring water samples of adjacent to the waste lagoon showed contamination with fecal bacteria and antibiotics resistance genes that probably originated from the waste lagoon (Koike et al. 2007). Although long-term passive storage of lagoon is an economical solution to decrease ARGs levels integrity of the lagoon walls should be maintained to ensure that ARGs from the lagoon do not contaminate groundwater. Further, presence of agricultural soils may serve as an exchange platform for genes with human (Fig. 6).

6 Environmental and Ecological Effects of Soil Antibiotics and Related ARB and ARGs

Environmental risks of most pharmaceuticals products were first identified in the 1990s followed by a series of monitoring and effect studies (Ågerstrand et al. 2015). Antibiotics, just like most pharmaceutical drugs, were designed to act effectively at low doses with short half-life and to be flushed out of the body through excretion (Thiele-Bruhn 2003). The waste matter can enter the environment through several pathways (Arun et al. 2017). These remains result in residual concentrations of antibiotics and increased abundance of antibiotic resistant microbes in the environment. Several studies on the environmental and ecological processes involved in the acquisition of resistance have been carried out (Aminov 2009; Berglund 2015; Grenni et al. 2018). However, these processes are complex resulting to limited data.

Previous studies report that antibiotics residues found in soil may have resulted from application of contaminated excrements on agricultural lands as fertilizers (Winckler and Grafe 2001). Most veterinary antibiotics are given to livestock animals as feed supplements and growth promoters, which are excreted as hazardous unmetabolized veterinary products (Ho et al. 2014). Continuous application of such excrements can increase quantities of antibiotics in soil which inhibit microbial growth or result to microbial resistance. These events promote growth of microbes thus changing the soil microbial community (Thiele-Bruhn 2003). Higher concentration of antibiotics or resistant bacteria can affect other organisms present in the soil. Moreover, they cause bioaccumulation of antibiotics in plants and contamination of nearby surface or groundwater (Ahmed et al. 2017; Kemper 2008). The risks posed by different antibiotic agents and related ARB and ARGs should be evaluated to reduce contamination of the surrounding environment.

7 Health Effects of Soil Contaminated Antibiotics and Their Related ARB and ARGs

A previous study reports acquisition of genes coding for resistance in soil bacteria (Séveno et al. 2002). In most cases resistance does not take place in the soil, but directly in the feces and then spreads to the soil through application of manure. Studies carried out in Sweden and the Netherlands on analysis of numerous fecal samples revealed high prevalence of bacteria resistant against various antibiotics (Iversen et al. 2002; Levy et al. 1988; Li et al. 2011). After application of manure, high level of tetracycline resistant bacteria are observed in soil and groundwater. However, the levels declined after cessation of slurry application to a level of non-slurry fertilized soil within 8 months (Nwosu 2001; Sengeløv et al. 2003). Human are directly exposed to soil either by direct contact or by alimentation chain as shown below (Fig. 7).

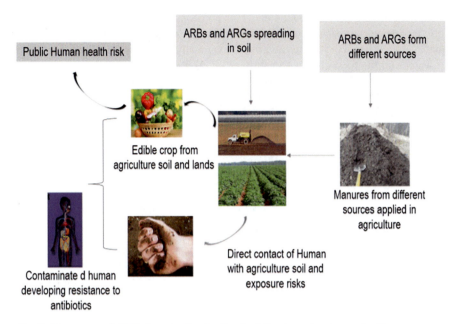

Fig. 7 Occurrence of ARGs in agricultural soil and relative impact on human health

7.1 Impact of Antibiotic Resistance on Human and Environmental Health

The overexposure to antibiotics increases the resistance levels of the human commensal microbiota (Austin and Anderson 1999). Enrichment in ARB is promoted by the presence of antimicrobials in the environment. Wastewater discharged from domestic sources can affect the diversity of ARB (Thevenon et al. 2012) and alters the genetic pool of water bodies by increasing abundance of ARGs. Hospital effluents have also been shown to be rich in ARB and ARGs (Zervos et al. 2003). The contamination of aquatic environments contributes to the spread of human pathogens and the dissemination of ARB and ARGs. In some aquatic systems, the cycle can include subsequent transmission of antibiotic resistance genes to bacteria associated with humans. Bacterial communities in aquatic systems comprise antibiotics producers and bacteria intrinsically resistant to several antibiotics. These two groups are natural carriers of genetic determinants of resistance.

Taking these factors together, aquatic environments can be seen as reactor system where drug resistance characteristics spread and recombine. Under such conditions, multidrug resistance features may emerge, and transmission to pathogenic bacteria is facilitated. Several studies have shown that in natural systems, the spread of antibiotic resistance is promoted. Environmental and pathogenic bacteria carry antibiotic resistance genes on mobile genetic elements and express them constitutively. Antibiotic resistance can reach the environment with the potential of adversely affecting aquatic and terrestrial organisms and might eventually reach humans through drinking water and the food chain (Aarestrup et al. 2008). The emergence of antibiotic resistance is a highly complex process and its significance of the interaction with bacterial populations is yet to be fully understood. The transfer of resistant bacteria to humans could occur via contaminated water or food. The transfer mechanism of antibiotic resistance from animals to humans still needs further investigation to better understand the transmission and contamination pathways.

7.2 Ecological Health Risk Related to the Antibiotics and the Antibiotic Resistance

Martínez et al. (2015) ecologically defined resistance as a function of ARGs to protect an organism from the inhibitory effect of an antimicrobial that is produced by another organism. This definition explains better the human health risks associated with antibiotics resistomes. The continuous use of antibiotics in agriculture has been described as a major contributor to the problem of resistant diseases in human medicine. Currently, antibiotics and their resistances are among the biggest threats to global health and food security. Several studies have highlighted the presence of antibiotics and antibiotic resistance in several environments unlikely to be contaminated with antibiotics used by humans.

The transmission of ARB and ARGs to humans is caused by ingestion of contaminated food or water and by direct contact with contaminated animals, soil, or water (Taylor et al. 2011). Resistant pathogens of agricultural origin get to humans through contact with livestock or through ingestion of bacteria from contaminated meat. Once a person is infected or colonized with a resistant microbe, human to human transmission is facilitated. The resistant microbe breaks through the species barrier and may be directly pathogenic to humans or a commensal with the ability to cause opportunistic infections. Resistance genes arising from the agricultural setting are introduced into human pathogens by horizontal gene transfer (HGT). The resulting resistant lineages are then selected for antibiotics use in humans. Comparison of a parental strain with a mutant strain or a strain containing a resistance gene acquired by HGT and introduced into human pathogens plays a major role in understanding human health risks associated with antibiotic resistomes.

8 Suggestions and Mitigation Strategies

The control of wastes is very important and must be implemented to reduce antibiotic resistance. This can be accomplished by providing a practical strategy with additional advantages of nutrient management in order to protect the soil environment. For instance, to reduce antibiotic resistance emanating from manure application, sediment erosion should be mitigated and use of animal manures be limited. Besides, control of surface runoff, lagoon spills, and seepage should be highly encouraged. Improved manure collection and increased storage capacity might limit surface runoff. There should be practical and proper application of manure to meet the crop demands. Composting and digestion of livestock waste can treat antibiotic residues. Aeration, watering, and turning of compost accelerates antibiotic decomposition of monensin, chlortetracycline, and tylosin. Besides, simple storage of manure stockpiles results in significant antibiotic degradation (Storteboom et al. 2007).

There is need for treatment of wastewater to minimize the spread of resistance. WWTPs may represent a critical point to control the global spread of antibiotics resistance. Thermophilic anaerobic sludge digestion appears particularly promising and may achieve superior ARGs removal relative to mesophilic digestion. More advanced treatment technologies such as membrane separation should be used to retain bacterial cells including their genetic material (Breazeal et al. 2013). In addition, ozone has been proposed to disinfect ARB and destroy ARGs.

Although there are trade-offs with air quality and cost of alternative fertilizers, incineration is a zero-risk solution with regard to reduction of antibiotics and ARGs. If used appropriately, incineration may be an alternative source of energy. Landfills still pose some risks because leachates may pollute groundwater and surface water as they are commonly redirected to a municipal WWTP (Renou et al. 2008). Hospital and industrial waste are considered as hotspots for antibiotics and antibiotic resistance spread in soil. Proper management of hospital wastes is of high concern. Membrane bioreactors should be used as targeted pretreatment systems to treat hospitals waste before discharging into public sewer systems since they can partially remove antibiotics and other drugs, as well as antibacterial resistance (Kovalova et al. 2012). Other potential sources of antibiotic resistance are pharmaceutical drug manufacturing sites (Fick et al. 2009; Kristiansson et al. 2011; Larsson et al. 2007; Li et al. 2009, 2010).

To define safe exposure levels is not possible in a strict sense. The scientific community should put effort to develop and provide regulators with a basis for defining and implementing standards. This would allow for various mitigation strategies. However, we must acknowledge that the uncertainty is still high regarding ultimate benefits for individual measures. At present, efficiency of the mitigation efforts can be best evaluated on the basis of surrogate measures such as the abundance of antibiotics and ARGs in the environment. Routine monitoring programs are required to provide baseline data that can help in assessing the conditions before and after mitigation activities. Establishing and maintaining existing biobanks of soil will allow for retrospective analyses.

Further treatments are also needed for the solid matrices. The maximal quantity of antibiotics residues in effluent of wastewater rejected in soil matrices should be controlled and regulations must be established accordingly. The same regulations must be implemented to control the quantity of antibiotics in sludge treatment plants (STPs) as well as in animal's manures before being used as soil fertilizer. Some rules should be set, for example, to reduce the use of sewage sludge or animal manures on agricultural land, thereby reducing the risks of spreading resistance in the environment to humans, food crops, and wildlife.

It is our responsibility to establish measures and major strategies for the prevention and control of antimicrobial resistances including exploiting the advantages of joint prevention and control. The investment in the research and development of antimicrobials and strengthening the management of antibacterial agents provides security to the environment. Optimization of antimicrobial consumption and resistance surveillance systems should be adopted to improve the capacity of professional personnel in antimicrobial resistance prevention and control. Furthermore, there is need to strengthening the prevention and management of environmental pollution by antimicrobials.

9 Conclusion

Conventionally, the presence of ARB and ARGs in soil is believed to stem from the exposure of animal and human bacteria from anthropogenic sources such as sewage and animal husbandry to selective pressure from clinically and agriculturally derived antibiotics compounds. The application of manures has been reported to be the main source of ARB and ARGs into soil. WWTPs also play a significant role in discharging the antibiotics residues and their related ARB and ARGs into soil. The end result is a generation of soil-borne ARB that facilitates spread of the ARGs in soil through HGT. The long-term application of manures, WWTP effluents, and wastewater are the main factors increasing and influencing the abundance and diversity of ARB and ARGs in soil. Moreover, manures origin and nature can effectively influence soil microbial community and the spreading of ARGs. The accelerated development and spread of ARB and ARGs in soil affects considerably the food chain supply causing severe health risks to humans. Very few effective policies both at national and international level on the use of antibiotics are available to deal with the actual challenge and control the ARB and ARGs pollution.

Acknowledgments This work was supported by the National Science Foundation of China (No. 31570504), the Natural Science Foundation of Tianjin (No. 16JCYBJC22900), and the Ministry of Education of China as an innovative team project (grant No. IRT13024).

Declaration The authors declare no conflict of interests regarding this paper.

Credit Authorship Contribution *Brim Stevy Ondon*: Conceptualization, Methodology, Software, Writing, review and editing – Original draft. *Shengnan Li*: Software and review. *Qixing Zhou*: Review and editing. *Fengxiang Li*: Supervision, Writing – review and editing, Project administration.

References

Aarestrup FM, Wegener HC, Collignon P (2008) Resistance in bacteria of the food chain: epidemiology and control strategies. Expert Rev Anti-Infect Ther 6(5):733–750

Ågerstrand M, Berg C, Björlenius B, Breitholtz M, Brunström Br, Fick J, Gunnarsson L, Larsson DJ, Sumpter JP, Tysklind M (2015) Improving environmental risk assessment of human pharmaceuticals. Environ Sci Technol 49(9):5336–5345

Ahmed S, Ibrahim M, Waheed R, Azdee ABH, Hashmi MZ, Ahmed S (2017) Genotoxicity and biochemical toxicity of soil antibiotics to earthworms. In: Antibiotics and antibiotics resistance genes in soils. Springer, Cham, pp 327–340

Almakki A, Jumas-Bilak E, Marchandin H, Licznar-Fajardo P (2019) Antibiotic resistance in urban runoff. Sci Total Environ 667:64–76

Aminov RI (2009) The role of antibiotics and antibiotic resistance in nature. Environ Microbiol 11 (12):2970–2988

Arun S, Mukhopadhyay M, Chakraborty P (2017) A review on antibiotics consumption, physico-chemical properties and their sources in Asian soil. In: Antibiotics and antibiotics resistance genes in soils. Springer, Cham, pp 39–53

Austin D, Anderson R (1999) Studies of antibiotic resistance within the patient, hospitals and the community using simple mathematical models. Philos Trans R Soc Lond B Biol Sci 354 (1384):721–738

Bassil RJ, Bashour II, Sleiman FT, Abou-Jawdeh YA (2013) Antibiotic uptake by plants from manure-amended soils. J Environ Sci Health B 48(7):570–574

Berglund B (2015) Environmental dissemination of antibiotic resistance genes and correlation to anthropogenic contamination with antibiotics. Infect Ecol Epidemiol 5(1):28564

Boxall AB, Fogg LA, Kay P, Blackwel PA, Pemberton EJ, Croxford A (2003a) Prioritisation of veterinary medicines in the UK environment. Toxicol Lett 142(3):207–218

Boxall AB, Kolpin DW, Halling-Sørensen B, Tolls J (2003b) Peer reviewed: are veterinary medicines causing environmental risks? ACS Publications, Washington

Breazeal MVR, Novak JT, Vikesland PJ, Pruden A (2013) Effect of wastewater colloids on membrane removal of antibiotic resistance genes. Water Res 47(1):130–140

Burkhardt M, Stamm C, Waul C, Singer H, Müller S (2005) Surface runoff and transport of sulfonamide antibiotics and tracers on manured grassland. J Environ Qual 34(4):1363–1371

Cao Y, Chang Z, Wang J, Ma Y, Fu G (2013) The fate of antagonistic microorganisms and antimicrobial substances during anaerobic digestion of pig and dairy manure. Bioresour Technol 136:664–671

Chee-Sanford JC, Aminov RI, Krapac I, Garrigues-Jeanjean N, Mackie RI (2001) Occurrence and diversity of tetracycline resistance genes in lagoons and groundwater underlying two swine production facilities. Appl Environ Microbiol 67(4):1494–1502

Chee-Sanford J, Mackie R, Koike S, Krapac I, Maxwell S, Lin YF, Aminov R (2009a) Fate and transport of antibiotic residues and antibiotic resistance genetic determinants during manure storage, treatment, and land application. J Environ Qual 38:1086–1108

Chee-Sanford JC, Mackie RI, Koike S, Krapac IG, Lin Y-F, Yannarell AC, Maxwell S, Aminov RI (2009b) Fate and transport of antibiotic residues and antibiotic resistance genes following land application of manure waste. J Environ Qual 38(3):1086–1108

Chen H, Zhang M (2013) Occurrence and removal of antibiotic resistance genes in municipal wastewater and rural domestic sewage treatment systems in eastern China. Environ Int 55:9–14

Chen B, Liang X, Huang X, Zhang T, Li X (2013) Differentiating anthropogenic impacts on ARGs in the Pearl River estuary by using suitable gene indicators. Water Res 47(8):2811–2820

Chen C, Li J, Chen P, Ding R, Zhang P, Li X (2014) Occurrence of antibiotics and antibiotic resistances in soils from wastewater irrigation areas in Beijing and Tianjin, China. Environ Pollut 193:94–101

Chen Q-L, An X-L, Li H, Zhu Y-G, Su J-Q, Cui L (2017) Do manure-borne or indigenous soil microorganisms influence the spread of antibiotic resistance genes in manured soil? Soil Biol Biochem 114:229–237

Chen C, Pankow CA, Oh M, Heath LS, Zhang L, Du P, Xia K, Pruden A (2019a) Effect of antibiotic use and composting on antibiotic resistance gene abundance and resistome risks of soils receiving manure-derived amendments. Environ Int 128:233–243

Chen Z, Zhang W, Yang L, Stedtfeld RD, Peng A, Gu C, Boyd SA, Li H (2019b) Antibiotic resistance genes and bacterial communities in cornfield and pasture soils receiving swine and dairy manures. Environ Pollut 248:947–957

D'Costa VM, King CE, Kalan L, Morar M, Sung WW, Schwarz C, Froese D, Zazula G, Calmels F, Debruyne R, Golding GB, Poinar HN, Wright GD (2011) Antibiotic resistance is ancient. Nature 477(7365):457–461

Davis J, Truman C, Kim S, Ascough J, Carlson K (2006) Antibiotic transport via runoff and soil loss. J Environ Qual 35(6):2250–2260

Ding GC, Radl V, Schloter-Hai B, Jechalke S, Heuer H, Smalla K, Schloter M (2014) Dynamics of soil bacterial communities in response to repeated application of manure containing sulfadiazine. PLoS One 9(3):e92958

Dungan RS, McKinney CW, Leytem AB (2018) Tracking antibiotic resistance genes in soil irrigated with dairy wastewater. Sci Total Environ 635:1477–1483

Fatta-Kassinos D, Kalavrouziotis I, Koukoulakis P, Vasquez M (2011a) The risks associated with wastewater reuse and xenobiotics in the agroecological environment. Sci Total Environ 409 (19):3555–3563

Fatta-Kassinos D, Meric S, Nikolaou A (2011b) Pharmaceutical residues in environmental waters and wastewater: current state of knowledge and future research. Anal Bioanal Chem 399 (1):251–275

Fick J, Söderström H, Lindberg RH, Phan C, Tysklind M, Larsson DJ (2009) Contamination of surface, ground, and drinking water from pharmaceutical production. Environ Toxicol Chem 28 (12):2522–2527

Forsberg KJ, Patel S, Gibson MK, Lauber CL, Knight R, Fierer N, Dantas G (2014a) Bacterial phylogeny structures soil resistomes across habitats. Nature 509(7502):612

Forsberg Z, Mackenzie AK, Sørlie M, Røhr ÅK, Helland R, Arvai AS, Vaaje-Kolstad G, Eijsink VG (2014b) Structural and functional characterization of a conserved pair of bacterial cellulose-oxidizing lytic polysaccharide monooxygenases. Proc Natl Acad Sci 111(23):8446–8451

Gao P, Mao D, Luo Y, Wang L, Xu B, Xu L (2012) Occurrence of sulfonamide and tetracycline-resistant bacteria and resistance genes in aquaculture environment. Water Res 46(7):2355–2364

Garner E, Benitez R, von Wagoner E, Sawyer R, Schaberg E, Hession WC, Krometis L-AH, Badgley BD, Pruden A (2017) Stormwater loadings of antibiotic resistance genes in an urban stream. Water Res 123:144–152

Gatica J, Cytryn E (2013) Impact of treated wastewater irrigation on antibiotic resistance in the soil microbiome. Environ Sci Pollut Res 20(6):3529–3538

Graham DW, Bergeron G, Bourassa MW, Dickson J, Gomes F, Howe A, Kahn LH, Morley PS, Scott HM, Simjee S (2019) Complexities in understanding antimicrobial resistance across domesticated animal, human, and environmental systems. Ann N Y Acad Sci 1441(1):17

Grenni P, Ancona V, Caracciolo AB (2018) Ecological effects of antibiotics on natural ecosystems: a review. Microchem J 136:25–39

Guo C, Wang K, Hou S, Wan L, Lv J, Zhang Y, Qu X, Chen S, Xu J (2017) H2O2 and/or TiO2 photocatalysis under UV irradiation for the removal of antibiotic resistant bacteria and their antibiotic resistance genes. J Hazard Mater 323:710–718

Hashmi MZ, Mahmood A, Kattel DB, Khan S, Hasnain A, Ahmed Z (2017) Antibiotics and antibiotic resistance genes (ARGs) in soil: occurrence, fate, and effects. In: Xenobiotics in the soil environment. Springer, Cham, pp 41–54

Ho YB, Zakaria MP, Latif PA, Saari N (2014) Environmental risk assessment for veterinary antibiotics and hormone in Malaysian agricultural soil. Iran J Public Health 43(Supple 3):67–71

Holten Lützhøft H-C, Vaes WH, Freidig AP, Halling-Sørensen B, Hermens JL (2000) Influence of pH and other modifying factors on the distribution behavior of 4-quinolones to solid phases and humic acids studied by "negligible-depletion" SPME– HPLC. Environ Sci Technol 34 (23):4989–4994

Hu X, Zhou Q, Luo Y (2010) Occurrence and source analysis of typical veterinary antibiotics in manure, soil, vegetables and groundwater from organic vegetable bases, northern China. Environ Pollut 158(9):2992–2998

Iversen A, Kühn I, Franklin A, Möllby R (2002) High prevalence of vancomycin-resistant enterococci in Swedish sewage. Appl Environ Microbiol 68(6):2838–2842

Ji X, Shen Q, Liu F, Ma J, Xu G, Wang Y, Wu M (2012) Antibiotic resistance gene abundances associated with antibiotics and heavy metals in animal manures and agricultural soils adjacent to feedlots in Shanghai; China. J Hazard Mater 235:178–185

Kay P, Blackwell PA, Boxall AB (2005) Column studies to investigate the fate of veterinary antibiotics in clay soils following slurry application to agricultural land. Chemosphere 60 (4):497–507

Kemper N (2008) Veterinary antibiotics in the aquatic and terrestrial environment. Ecol Indic 8 (1):1–13

Knapp CW, Dolfing J, Ehlert PA, Graham DW (2010a) Evidence of increasing antibiotic resistance gene abundances in archived soils since 1940. Environ Sci Technol 44(2):580–587

Knapp CW, Zhang W, Sturm BS, Graham DW (2010b) Differential fate of erythromycin and beta-lactam resistance genes from swine lagoon waste under different aquatic conditions. Environ Pollut 158(5):1506–1512

Knapp CW, McCluskey SM, Singh BK, Campbell CD, Hudson G, Graham DW (2011) Antibiotic resistance gene abundances correlate with metal and geochemical conditions in archived Scottish soils. PLoS One 6(11):e27300

Knapp CW, Callan AC, Aitken B, Shearn R, Koenders A, Hinwood A (2017) Relationship between antibiotic resistance genes and metals in residential soil samples from Western Australia. Environ Sci Pollut Res 24(3):2484–2494

Koike S, Krapac I, Oliver H, Yannarell A, Chee-Sanford J, Aminov R, Mackie RI (2007) Monitoring and source tracking of tetracycline resistance genes in lagoons and groundwater adjacent to swine production facilities over a 3-year period. Appl Environ Microbiol 73 (15):4813–4823

Kovalova L, Siegrist H, Singer H, Wittmer A, McArdell CS (2012) Hospital wastewater treatment by membrane bioreactor: performance and efficiency for organic micropollutant elimination. Environ Sci Technol 46(3):1536–1545

Kreuzig R, Höltge S (2005) Investigations on the fate of sulfadiazine in manured soil: laboratory experiments and test plot studies. Environ Toxicol Chem Int J 24(4):771–776

Kristiansson E, Fick J, Janzon A, Grabic R, Rutgersson C, Weijdegård B, Söderström H, Larsson DJ (2011) Pyrosequencing of antibiotic-contaminated river sediments reveals high levels of resistance and gene transfer elements. PLoS One 6(2):e17038

Kumar K, Gupta SC, Chander Y, Singh AK (2005) Antibiotic use in agriculture and its impact on the terrestrial environment. Adv Agron 87:1–54

Kurwadkar ST, Adams CD, Meyer MT, Kolpin DW (2007) Effects of sorbate speciation on sorption of selected sulfonamides in three loamy soils. J Agric Food Chem 55(4):1370–1376

LaPara TM, Burch TR, McNamara PJ, Tan DT, Yan M, Eichmiller JJ (2011) Tertiary-treated municipal wastewater is a significant point source of antibiotic resistance genes into Duluth-Superior Harbor. Environ Sci Technol 45(22):9543–9549

Larsson DJ, de Pedro C, Paxeus N (2007) Effluent from drug manufactures contains extremely high levels of pharmaceuticals. J Hazard Mater 148(3):751–755

Latare A, Kumar O, Singh S, Gupta A (2014) Direct and residual effect of sewage sludge on yield, heavy metals content and soil fertility under rice–wheat system. Ecol Eng 69:17–24

Lebreton F, Manson AL, Saavedra JT, Straub TJ, Earl AM, Gilmore MS (2017) Tracing the enterococci from Paleozoic origins to the hospital. Cell 169(5):849–861.e13

Lee S, Suits M, Wituszynski D, Winston R, Martin J, Lee J (2020) Residential urban stormwater runoff: a comprehensive profile of microbiome and antibiotic resistance. Sci Total Environ 723:138033

Levy S, Marshall B, Schluederberg S, Rowse D, Davis J (1988) High frequency of antimicrobial resistance in human fecal flora. Antimicrob Agents Chemother 32(12):1801–1806

Li D, Yang M, Hu J, Zhang J, Liu R, Gu X, Zhang Y, Wang Z (2009) Antibiotic-resistance profile in environmental bacteria isolated from penicillin production wastewater treatment plant and the receiving river. Environ Microbiol 11(6):1506–1517

Li D, Yu T, Zhang Y, Yang M, Li Z, Liu M, Qi R (2010) Antibiotic resistance characteristics of environmental bacteria from an oxytetracycline production wastewater treatment plant and the receiving river. Appl Environ Microbiol 76(11):3444–3451

Li B, Sun J-Y, Liu Q-Z, Han L-Z, Huang X-H, Ni Y-X (2011) High prevalence of CTX-M β-lactamases in faecal *Escherichia coli* strains from healthy humans in Fuzhou, China. Scand J Infect Dis 43(3):170–174

Li N, Sheng G-P, Lu Y-Z, Zeng RJ, Yu H-Q (2017) Removal of antibiotic resistance genes from wastewater treatment plant effluent by coagulation. Water Res 111:204–212

Liu H, Zhou X, Huang H, Zhang J (2019) Prevalence of antibiotic resistance genes and their association with antibiotics in a wastewater treatment plant: process distribution and analysis. Water 11(12):2495

Lloret E, Pascual JA, Brodie EL, Bouskill NJ, Insam H, Juárez MF-D, Goberna M (2016) Sewage sludge addition modifies soil microbial communities and plant performance depending on the sludge stabilization process. Appl Soil Ecol 101:37–46

Looft T, Johnson TA, Allen HK, Bayles DO, Alt DP, Stedtfeld RD, Sul WJ, Stedtfeld TM, Chai B, Cole JR (2012) In-feed antibiotic effects on the swine intestinal microbiome. Proc Natl Acad Sci 109(5):1691–1696

Lucas D, Badia-Fabregat M, Vicent T, Caminal G, Rodríguez-Mozaz S, Balcázar J, Barceló D (2016) Fungal treatment for the removal of antibiotics and antibiotic resistance genes in veterinary hospital wastewater. Chemosphere 152:301–308

Luo Y, Mao D, Rysz M, Zhou Q, Zhang H, Xu L, Alvarez JJ, P. (2010) Trends in antibiotic resistance genes occurrence in the Haihe River, China. Environ Sci Technol 44(19):7220–7225

Martín J, Santos JL, Aparicio I, Alonso E (2015) Pharmaceutically active compounds in sludge stabilization treatments: anaerobic and aerobic digestion, wastewater stabilization ponds and composting. Sci Total Environ 503:97–104

Martínez JL, Coque TM, Baquero F (2015) What is a resistance gene? Ranking risk in resistomes. Nat Rev Microbiol 13(2):116–123

McGrath JW, Hammerschmidt F, Quinn JP (1998) Biodegradation of phosphonomycin by rhizobium huakuii PMY1. Appl Environ Microbiol 64(1):356–358

McKinney CW, Loftin KA, Meyer MT, Davis JG, Pruden A (2010) Tet and sul antibiotic resistance genes in livestock lagoons of various operation type, configuration, and antibiotic occurrence. Environ Sci Technol 44(16):6102–6109

McLain JE, Williams CF (2010) Development of antibiotic resistance in bacteria of soils irrigated with reclaimed wastewater. In: 5th National Decennial Irrigation Conference Proceedings, 5–8 December 2010, Phoenix convention Center, Phoenix, Arizona USA. American Society of Agricultural and Biological Engineers, Michigan, p 1

Michael I, Rizzo L, McArdell C, Manaia C, Merlin C, Schwartz T, Dagot C, Fatta-Kassinos D (2013) Urban wastewater treatment plants as hotspots for the release of antibiotics in the environment: a review. Water Res 47(3):957–995

Michelini L, Gallina G, Capolongo F, Ghisi R (2014) Accumulation and response of willow plants exposed to environmental relevant sulfonamide concentrations. Int J Phytoremediation 16(9):947–961

Munir M, Xagoraraki I (2011) Levels of antibiotic resistance genes in manure, biosolids, and fertilized soil. J Environ Qual 40(1):248–255

Munir M, Wong K, Xagoraraki I (2011) Release of antibiotic resistant bacteria and genes in the effluent and biosolids of five wastewater utilities in Michigan. Water Res 45(2):681–693

Nathan SS, Chung PG, Murugan K (2004) Effect of botanical insecticides and bacterial toxins on the gut enzyme of the rice leaffolderCnaphalocrocis medinalis. Phytoparasitica 32(5):433

Negreanu Y, Pasternak Z, Jurkevitch E, Cytryn E (2012) Impact of treated wastewater irrigation on antibiotic resistance in agricultural soils. Environ Sci Technol 46(9):4800–4808

Nesme J, Simonet P (2015) The soil resistome: a critical review on antibiotic resistance origins, ecology and dissemination potential in telluric bacteria. Environ Microbiol 17(4):913–930

Nesme J, Topp E, Simonet P (2015) Antibiotic resistance genes mobility and dissemination during a long-term experiment: fifteen years of farm-soil amendment with antibiotics analyzed with high-throughput sequencing. Indo-French Workshop on Environmental Biotechnology (Molecular Ecology & Environmental Engineering)

Novo A, Manaia CM (2010) Factors influencing antibiotic resistance burden in municipal wastewater treatment plants. Appl Microbiol Biotechnol 87(3):1157–1166

Nwosu VC (2001) Antibiotic resistance with particular reference to soil microorganisms. Res Microbiol 152(5):421–430

Oliver JP, Gooch CA, Lansing S, Schueler J, Hurst JJ, Sassoubre L, Crossette EM, Aga DS (2020) Invited review: fate of antibiotic residues, antibiotic-resistant bacteria, and antibiotic resistance genes in US dairy manure management systems. J Dairy Sci 103(2):1051–1071

Pan M, Chu L (2018) Occurrence of antibiotics and antibiotic resistance genes in soils from wastewater irrigation areas in the Pearl River Delta region, southern China. Sci Total Environ 624:145–152

Pehrsson EC, Tsukayama P, Patel S, Mejía-Bautista M, Sosa-Soto G, Navarrete KM, Calderon M, Cabrera L, Hoyos-Arango W, Bertoli MT (2016a) Interconnected microbiomes and resistomes in low-income human habitats. Nature 533(7602):212

Pehrsson SJ, Eglington BM, Evans DA, Huston D, Reddy SM (2016b) Metallogeny and its link to orogenic style during the Nuna supercontinent cycle. Geol Soc Lond, Spec Publ 424(1):83–94

Peng S, Feng Y, Wang Y, Guo X, Chu H, Lin X (2017) Prevalence of antibiotic resistance genes in soils after continually applied with different manure for 30 years. J Hazard Mater 340:16–25

Pepper IL, Brooks JP, Gerba CP (2018) Antibiotic resistant bacteria in municipal wastes: is there reason for concern? Environ Sci Technol 52(7):3949–3959

Perry J, Wright G (2013) The antibiotic resistance "mobilome": searching for the link between environment and clinic. Front Microbiol 4:138

Perry JA, Westman EL, Wright GD (2014) The antibiotic resistome: what's new? Curr Opin Microbiol 21:45–50

Petrie B, Barden R, Kasprzyk-Hordern B (2015) A review on emerging contaminants in wastewaters and the environment: current knowledge, understudied areas and recommendations for future monitoring. Water Res 72:3–27

Pruden A, Pei R, Storteboom H, Carlson KH (2006) Antibiotic resistance genes as emerging contaminants: studies in northern Colorado. Environ Sci Technol 40(23):7445–7450

Pruden A, Larsson DG, Amézquita A, Collignon P, Brandt KK, Graham DW, Lazorchak JM, Suzuki S, Silley P, Snape JR, Topp E, Zhang T, Zhu YG (2013) Management options for reducing the release of antibiotics and antibiotic resistance genes to the environment. Environ Health Perspect 121(8):878–885

Qian X, Gu J, Sun W, Wang X-J, Su J-Q, Stedfeld R (2018) Diversity, abundance, and persistence of antibiotic resistance genes in various types of animal manure following industrial composting. J Hazard Mater 344:716–722

Renou S, Givaudan J, Poulain S, Dirassouyan F, Moulin P (2008) Landfill leachate treatment: review and opportunity. J Hazard Mater 150(3):468–493

Rizzo L, Manaia C, Merlin C, Schwartz T, Dagot C, Ploy MC, Michael I, Fatta-Kassinos D (2013) Urban wastewater treatment plants as hotspots for antibiotic resistant bacteria and genes spread into the environment: a review. Sci Total Environ 447:345–360

Roig N, Sierra J, Martí E, Nadal M, Schuhmacher M, Domingo JL (2012) Long-term amendment of Spanish soils with sewage sludge: effects on soil functioning. Agric Ecosyst Environ 158:41–48

Sabri N, Schmitt H, Van der Zaan B, Gerritsen H, Zuidema T, Rijnaarts H, Langenhoff A (2020) Prevalence of antibiotics and antibiotic resistance genes in a wastewater effluent-receiving river in the Netherlands. J Environ Chem Eng 8(1):102245

Sengeløv G, Agersø Y, Halling-Sørensen B, Baloda SB, Andersen JS, Jensen LB (2003) Bacterial antibiotic resistance levels in Danish farmland as a result of treatment with pig manure slurry. Environ Int 28(7):587–595

Séveno NA, Kallifidas D, Smalla K, van Elsas JD, Collard J-M, Karagouni AD, Wellington EM (2002) Occurrence and reservoirs of antibiotic resistance genes in the environment. Rev Med Microbiol 13(1):15–27

Singh RP, Singh P, Ibrahim MH, Hashim R (2012) Land application of sewage sludge: physico-chemical and microbial response. In: Reviews of environmental contamination and toxicology. Springer, Cham, pp 41–61

Smith MS, Yang RK, Knapp CW, Niu Y, Peak N, Hanfelt MM, Galland JC, Graham DW (2004) Quantification of tetracycline resistance genes in feedlot lagoons by real-time PCR. Appl Environ Microbiol 70(12):7372–7377

Storteboom HN, Kim S-C, Doesken KC, Carlson KH, Davis JG, Pruden A (2007) Response of antibiotics and resistance genes to high-intensity and low-intensity manure management. J Environ Qual 36(6):1695–1703

Storteboom H, Arabi M, Davis J, Crimi B, Pruden A (2010) Identification of antibiotic-resistance-gene molecular signatures suitable as tracers of pristine river, urban, and agricultural sources. Environ Sci Technol 44(6):1947–1953

Sui Q, Zhang J, Chen M, Tong J, Wang R, Wei Y (2016) Distribution of antibiotic resistance genes (ARGs) in anaerobic digestion and land application of swine wastewater. Environ Pollut 213:751–759

Taylor V-J, Verner-Jeffreys DW, Baker-Austin C (2011) Aquatic systems: maintaining, mixing and mobilising antimicrobial resistance. Trends Ecol Evol 26:278–284

Thevenon F, Adatte T, Wildi W, Poté J (2012) Antibiotic resistant bacteria/genes dissemination in lacustrine sediments highly increased following cultural eutrophication of Lake Geneva (Switzerland). Chemosphere 86(5):468–476

Thiele-Bruhn S (2003) Pharmaceutical antibiotic compounds in soils–a review. J Plant Nutr Soil Sci 166(2):145–167

Urra J, Alkorta I, Mijangos I, Epelde L, Garbisu C (2019) Application of sewage sludge to agricultural soil increases the abundance of antibiotic resistance genes without altering the composition of prokaryotic communities. Sci Total Environ 647:1410–1420

Vaz-Moreira I, Varela AR, Pereira TV, Fochat RC, Manaia CM (2016) Multidrug resistance in quinolone-resistant gram-negative bacteria isolated from hospital effluent and the municipal wastewater treatment plant. Microb Drug Resist 22(2):155–163

Wallace JS, Garner E, Pruden A, Aga DS (2018) Occurrence and transformation of veterinary antibiotics and antibiotic resistance genes in dairy manure treated by advanced anaerobic digestion and conventional treatment methods. Environ Pollut 236:764–772

Wang FH, Qiao M, Chen Z, Su JQ, Zhu YG (2015) Antibiotic resistance genes in manure-amended soil and vegetables at harvest. J Hazard Mater 299:215–221

Wang F, Stedtfeld RD, Kim O-S, Chai B, Yang L, Stedtfeld TM, Hong SG, Kim D, Lim HS, Hashsham SA (2016) Influence of soil characteristics and proximity to Antarctic research stations on abundance of antibiotic resistance genes in soils. Environ Sci Technol 50(23):12621–12629

Wang M, Shen W, Yan L, Wang X-H, Xu H (2017) Stepwise impact of urban wastewater treatment on the bacterial community structure, antibiotic contents, and prevalence of antimicrobial resistance. Environ Pollut 231:1578–1585

Wang M, Liu P, Xiong W, Zhou Q, Wangxiao J, Zeng Z, Sun Y (2018) Fate of potential indicator antimicrobial resistance genes (ARGs) and bacterial community diversity in simulated manure-soil microcosms. Ecotoxicol Environ Saf 147:817–823

Wen X, Mi J, Wang Y, Ma B, Zou Y, Liao X, Liang JB, Wu Y (2019) Occurrence and contamination profiles of antibiotic resistance genes from swine manure to receiving environments in Guangdong Province southern China. Ecotoxicol Environ Saf 173:96–102

Wenzel RP, Edmond MB (2000) Managing antibiotic resistance. Mass Medical Society, Waltham

Winckler C, Grafe A (2001) Use of veterinary drugs in intensive animal production. J Soils Sediments 1(2):66

Wu N, Qiao M, Zhang B, Cheng W-D, Zhu Y-G (2010) Abundance and diversity of tetracycline resistance genes in soils adjacent to representative swine feedlots in China. Environ Sci Technol 44(18):6933–6939

Yang Y, Li B, Zou S, Fang HH, Zhang T (2014) Fate of antibiotic resistance genes in sewage treatment plant revealed by metagenomic approach. Water Res 62:97–106

Zervos MJ, Hershberger E, Nicolau DP, Ritchie DJ, Blackner LK, Coyle EA, Donnelly AJ, Eckel SF, Eng RH, Hiltz A (2003) Relationship between fluoroquinolone use and changes in susceptibility to fluoroquinolones of selected pathogens in 10 United States teaching hospitals, 1991–2000. Clin Infect Dis 37(12):1643–1648

Zhang T, Yang Y, Pruden A (2015) Effect of temperature on removal of antibiotic resistance genes by anaerobic digestion of activated sludge revealed by metagenomic approach. Appl Microbiol Biotechnol 99(18):7771–7779

Zhang C, Li Y, Wang C, Niu L, Cai W (2016) Occurrence of endocrine disrupting compounds in aqueous environment and their bacterial degradation: a review. Crit Rev Environ Sci Technol 46(1):1–59

Zhao X, Wang J, Zhu L, Ge W, Wang J (2017) Environmental analysis of typical antibiotic-resistant bacteria and ARGs in farmland soil chronically fertilized with chicken manure. Sci Total Environ 593:10–17

Zhou Y, Niu L, Zhu S, Lu H, Liu W (2017) Occurrence, abundance, and distribution of sulfonamide and tetracycline resistance genes in agricultural soils across China. Sci Total Environ 599:1977–1983

Zhu Y-G, Johnson TA, Su J-Q, Qiao M, Guo G-X, Stedtfeld RD, Hashsham SA, Tiedje JM (2013) Diverse and abundant antibiotic resistance genes in Chinese swine farms. Proc Natl Acad Sci 110(9):3435–3440

An Overview of Morpho-Physiological, Biochemical, and Molecular Responses of Sorghum Towards Heavy Metal Stress

Dewanshi Mishra, Smita Kumar, and Bhartendu Nath Mishra

Contents

1	Introduction	156
2	Heavy Metal Uptake, Translocation, and Accumulation in Sorghum	158
3	Impact of Heavy Metals on the Morphological Characteristics of Sorghum	161
4	Effect of Heavy Metals on the Physiological Aspects of Sorghum	162
5	Biochemical Changes in Response to Heavy Metals in Sorghum	164
6	Molecular Modulations Under Heavy Metal Stress in Sorghum	165
7	Future Prospects	170
8	Conclusion	170
References		171

Abstract Heavy metal (HM) contamination is a serious global environmental crisis. Over the past decade, industrial effluents, modern agricultural practices, and other anthropogenic activities have significantly depleted the soil environment. In plants, metal toxicity leads to compromised growth, development, productivity, and yield. Also, HMs negatively affect human health due to food chain contamination. Thus, it is imperative to reduce metal accumulation and toxicity. In nature, certain plant species exhibit an inherent capacity of amassing large amounts of HMs with remarkable tolerance. These plants with unique characteristics can be employed for the remediation of contaminated soil and water. Among different plant species, *Sorghum bicolor* has the potential of accumulating huge amounts of HMs, thus could be regarded as a hyperaccumulator. This means that it is a metal tolerant, high

---

Dewanshi Mishra and Smita Kumar contributed equally to this work.

D. Mishra · B. N. Mishra
Department of Biotechnology, Institute of Engineering and Technology, Dr. A.P.J. Abdul Kalam Technical University, Lucknow, Uttar Pradesh, India
e-mail: dewanshimishra17@gmail.com; profbnmishra@gmail.com

S. Kumar (✉)
Department of Biochemistry, King George's Medical University, Lucknow, Uttar Pradesh, India
e-mail: smitabiochem@gmail.com

© The Author(s), under exclusive license to Springer Nature Switzerland AG 2020
P. de Voogt (ed.), *Reviews of Environmental Contamination and Toxicology Volume 256*, Reviews of Environmental Contamination and Toxicology 256,
https://doi.org/10.1007/398_2020_61

biomass producing energy crop, and thus can be utilized for phytoremediation. However, high concentrations of HMs hamper plant height, root hair density, shoot biomass, number of leaves, chlorophyll, carotenoid, and carbohydrate content. Thus, understanding the response of Sorghum towards different HMs holds considerable importance. Considering this, we have uncovered the basic information about the metal uptake, translocation, and accumulation in Sorghum. Plants respond to different HMs via sensing, signaling, and modulations in physico-chemical processes. Therefore, in this review, a glimpse of HM toxicity and the response of Sorghum at the morphological, physiological, biochemical, and molecular levels has been provided. The review highlights the future research needs and emphasizes the extensive molecular dissection of Sorghum to explore its genetic adaptability towards different abiotic stresses that can be exploited to develop resilient crop varieties.

Keywords Antioxidant · Biochemical · Heavy metals · Hyperaccumulation · Oxidative stress · Phytoremediation · *Sorghum bicolor*

Abbreviations

ACP	Acid phosphatase
APX	Ascorbate peroxidase
CAT	Catalase
CAX	Cation exchanger
CDF	Cation diffusion facilitator
CTR	Copper transporter
DHA	Dehydroascorbic acid
DHAR	Dehydroascorbate reductase
GR	Glutathione reductase
GSH	Glutathione
GSSG	Glutathione disulfide
HM	Heavy metal
MDA	Malondialdehyde
MDHAR	Mono-dehydroascorbate reductase
POX	Peroxidase
QTL	Quantitative trait locus
ROS	Reactive oxygen species
SOD	Superoxide dismutase
ZIP	Zinc-iron permease

1 Introduction

Rapidly growing industrialization and urbanization are a concern due to the release and contamination of heavy metals (HMs) in the environment through toxic effluents. Also, global nuclear weapon production has released highly toxic wastes containing radionuclides such as Uranium (U), Cesium (Cs), and Strontium (Sr) into the groundwater and subsurface soils. As a result, HM toxicity has severely influenced the ecosystem, especially, impeded plant growth, development, and yield. Besides, HM toxicity has deteriorated human health due to food chain contamination (Rai et al. 2019; Masindi and Muedi 2018). Therefore, the sustainable production of crops is imperative for both humans and the environment. Studies have demonstrated that HMs exert toxicity via competing with the essential elements for the entry into the root cell, interacting with the sulfhydryl groups (-SH) of functional proteins and inducing oxidative stress in plants causing ionic and cellular redox imbalance (Demecsová and Tamás 2019; Kumar and Trivedi 2015a; Mittler 2002).

Few plant species have a natural capacity of accumulating large amounts of HMs and are known as hyperaccumulators. The hyperaccumulating plant species have the potential to be utilized for natural plant-based remediation, known as phytoremediation. One such metal accumulating plant is a cereal-grass called Sorghum, a member of the Poaceae, considered an important plant family for phytoremediation and biofuel production (Soudek et al. 2014; Yaashikaa et al. 2019; Rekik et al. 2017; Al Chami et al. 2015; Shi-Qi et al. 2018; Wang et al. 2016). Sorghum is the fifth most cultivated crop across the globe after wheat, maize, rice, and barley. Also, it is a staple crop for a large population living in arid and semi-arid regions. Sorghum possesses a deep and fibrous root system, primary culm, and the capacity for both basal and axillary tillering (Rooney 2014). Until date, all the cultivated Sorghum species are a variety of *Sorghum bicolor* subspecies *arundinaceum*. The most cultivated varieties are *Sorghum bicolor*, *Sorghum drummondii*, and *Sorghum sudanense* (Paterson et al. 2009; Rooney 2014).

Sorghum bicolor (S. bicolor) is commonly used for human food, animal feed, and ethanol production. Being the domesticated variety, the Sorghum cultivar improvement depends upon various traits like tillering, regrowth, plant height, stem type, maturity, and grain yield potential. The variation in these traits categorizes them into grain Sorghum, forage Sorghum, and energy Sorghum (Rooney 2014; Zheng et al. 2011). It is considered as one of the crops with the potential to survive against future environmental challenges as it has an intuitive ability to adapt to conditions such as drought, salinity, and high temperatures (Abou-Elwafa and Shehzad 2018; Almodares et al. 2014; Chopra et al. 2017; Phuong et al. 2019).

All plants typically manifest sensitivity/tolerance/bioaccumulation upon exposure to an array of HMs. Interestingly, a perennial bunchgrass of the family Poaceae, *Chrysopogon zizanioides*, possesses phytostabilization potential against different HMs such as Iron (Fe), Manganese (Mn), Zinc (Zn), Copper (Cu), Lead (Pb), Nickel (Ni), and Chromium (Cr), with no DNA damage response or genetic instability (Banerjee et al. 2016). Likewise, *Salsola soda* and *Cirsium arvense* have been identified to have phytoextraction and phytostabilization potential (Lorestani et al. 2013). Among different plant species, black gram and Sorghum demonstrate phytoextraction capacities (Padmapriya et al. 2016). *S. bicolor* has high bioaccumulation, phytoremediation, and phytostabilization capacities against Cadmium (Cd), Zn (Soudek et al. 2014; Wang et al. 2017a), Cr (Yaashikaa et al. 2019),

Cu, Ni, Pb (Al Chami et al. 2015), U, Sr (Wang et al. 2017b), and Cs (Wang et al. 2016, 2017c). Symptomatically, the HM stress profoundly influences morphological and biochemical parameters of crop plants like millet (*Eleusine coracana*), mustard (*Brassica juncea*), black gram (*Vigna mungo*), and pumpkin (*Telfairia occidentalis*) (Padmapriya et al. 2016). Therefore, an insight into the morpho-physiological, biochemical, and molecular responses of Sorghum towards HM stress will complement the possibility of utilizing *S. bicolor* for phytoremediation including phytoextraction, bioaccumulation, and phytostabilization.

To negate HM toxicity, plants require complex mechanisms at the cellular, tissue, biochemical, physiological, and molecular levels. In the past decade, the "omics" studies have extended our understanding and knowledge related to the expression of stress-responsive genes, transcription factors, proteins, and metabolites in plants. Transcriptome, proteome, metabolome, and ionome profiling have been carried out in different plants in response to HMs at various doses (Kumar et al. 2015; Kumar and Trivedi 2019; Shukla et al. 2018; Xie et al. 2015). Collectively, these studies have identified numerous stress-responsive genes, heat-shock proteins, transcription factors, metallothioneins, signaling protein kinases, transporters, and antioxidant enzymes having an intricate role in detoxification of metal ions (Chakrabarty et al. 2009; Dixit et al. 2016; Kumar et al. 2011, 2013a, b). Different functional genomics approaches including ectopic expression of stress-responsive genes have established possibilities of resilience towards different abiotic stresses including HMs (Kumar et al. 2019; Kumar et al. 2013a, b; Dutta et al. 2018; Tang et al. 2019). Through the collation of the information generated through omic studies, the possibility of developing stress-tolerant varieties by breeding or genetic engineering can be broadened. Besides, the advantage of omic studies on Sorghum will extend to other grass species that might be further identified as potential resources for phytoremediation. In this review, the focus is on the response of Sorghum towards different HMs at the morphological, biochemical, physiological, and molecular levels. Furthermore, it emphasizes the extensive molecular dissection of Sorghum to explore its genetic adaptability towards different abiotic stresses that can be exploited to develop resilient crop varieties.

2 Heavy Metal Uptake, Translocation, and Accumulation in Sorghum

The uptake, transport, and utilization pathways of HMs are under tight regulation in plant cells. All plants tend to accumulate HMs to a certain level in their tissues regardless of their sensitivity/tolerance/phytoremediation capabilities. Plants uptake HMs through their root systems while various ion channels assist in their translocation to the plant tissues. A growing knowledge of metal transporters such as Cation Diffusion Facilitator (CDF), Cation exchanger (CAX), Copper Transporter (CTR), Heavy Metal ATPases (HM-ATPase), and Zinc-Iron Permease (ZIP) have helped in

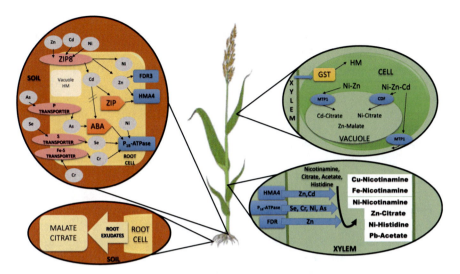

Fig. 1 Heavy metal uptake, translocation, and detoxification. *ABA* abscisic acid pathway, *CDF* cation diffusion facilitator, *FDR3* a multidrug and toxin efflux family molecule, *Fe-S* iron sulfur transporter protein, *GST* glutathione-S-transferase, *HM* heavy metal, *HMA4* heavy metal ATPase 4, *MTP1* metal transporter protein 1, *P1B-ATPases* heavy metal ATPases, *ZIP8* zinc-iron permease, *ZIP* zinc-iron protein pathway

understanding the uptake mechanism in plants (Cong et al. 2019; Migeon et al. 2010). In this regard, the mechanistic experiments, and the molecular and genomic based studies on the transporters responsible for metal uptake and accumulation in Sorghum need to be investigated in detail.

In terms of metal uptake, studies have suggested that the physico-chemical properties of soil such as density, pH, organic matter, and moisture content influence the bioavailability of metals in distinct ways (Adamczyk-Szabela et al. 2015; Chibuike and Obiora 2014). For example, in acidic soils, uptake of Cd increases in plants (Melo et al. 2014). The organic matter decreases Cd concentration in roots by decreasing the bioavailability in soil and increases Cd translocation in shoots (Pinto et al. 2005). Also, organic matter is reported to decrease the uptake of Cu, Zn, and Fe in *S. bicolor* (Pinto et al. 2004).

After metal enters into the root cell, it is transported to the xylem vessels and translocated to the above-ground parts (Fig. 1). In this context, Casparian strips in the endodermis act as a diffusion-barrier and regulate the apoplastic pathway. These strips are composed of lignin and suberin and play an important role in the translocation of metals in plants. For instance, Casparian strips in the maize root restrict the entry of Cd and Pb in the central cylinder and limit the translocation into the shoots (Seregin et al. 2004). Recently, the investigation of two contrasting genotypes of *S. bicolor* has revealed that the formation of Casparian strip influences the differential uptake and translocation of Cd (Feng et al. 2018). Moreover, the suberized Casparian band blocks the flow of HMs, and these metals accumulate in the root cells (Ghori et al. 2019). It has been ascertained that ethylene diamine tetraacetic acid (EDTA) forms a complex with the metal and is taken up by the roots through the apoplastic pathway. This EDTA disrupts the Casparian strip and facilitates the translocation of metal in the shoot. Thus, disruption of these Casparian strips helps in the phytoextraction of HMs (Nowack et al. 2008; Shahid et al. 2014). Notably, several anatomical changes occur in the root structure during the uptake and translocation of HMs. Cd and Cu stress lead to the remarkable narrowing of the metaxylem vessels in different tissues of *S. bicolor* (Kasim 2006). It is noteworthy that the in-depth investigation of the structure and function dynamics of the Casparian strip shall help in elucidation of the effect of different HMs on the radial transport properties of roots.

The Translocation Factor (TF), referred to as the shoot-root quotient determines the concentration of metal in the aerial parts of the plant translocated from the root tissue (Singh et al. 2010). The investigation of 96 genotypes of *S. bicolor* revealed TF of Cd between 0.052 and 0.22 (Jia et al. 2017). Also, an increased uptake and accumulation of Cu and Fe with high Transfer Coefficient (TC) for Mo and Pb has been observed in *S. bicolor* (Galavi et al. 2010). Higher accumulation potential for Cr (VI) in roots has been reported in Sorghum in comparison to stem and leaves (Yaashikaa et al. 2019). A dose-dependent accumulation of Cr has been found in the seedlings of *S. bicolor* with accumulation in the order of roots>leaves>shoots (Shanker and Pathmanabhan 2004). In five different cultivars of *S. bicolor,* Cd and Zn concentration has been reported to be much higher in roots than shoots. However, high concentrations of these HMs promoted their root-to-shoot translocation (Soudek et al. 2013). Sorghum cultivars, Keller, Mray, and Rio have been identified to accumulate high Pb concentrations in leaves; Cd, Cu, and Zn concentrations in stem while soil treatment with NH_4NO_3 and $(NH_4)_2SO_4$ increased Zn and Cd accumulation in roots (Zhuang et al. 2009). A study demonstrated that Cr is not absorbed directly but transported and accumulated via carrier ions such as sulfate or iron in *S. bicolor* (Gajalakshmi et al. 2012).

It has been suggested that the HM translocation and accumulation from the root to shoot in a plant increases with an increase in age (Diwan et al. 2012). The differential accumulation potential of different HMs in roots and shoots of *S. bicolor* has been ascertained (Marchiol et al. 2007). For example, *S. bicolor* shoot accumulates 16, 28, and 17 times less Arsenic (As) in the shoot in comparison to roots at 14 days of exposure to 6.7, 33.5, and 67 µM As concentrations, respectively (Shaibur et al. 2008). Also, *S. bicolor* has been reported to accumulate Cr, Cd, and Pb above the Critical Permissible Limits for fodder in addition to Co and Cu (Bhatti et al. 2016). The metal Bioaccumulation Factor (BAF) is an indicator of the plant capacity to uptake and accumulate HMs in tissues (Oh et al. 2015). The value of BAF above 1 is considered extremely harmful for the organisms (Singh et al. 2010). In *S. bicolor*, the values of BAF for Cd, Cr, and Pb have been reported to be 1 and above 1, respectively (Bhatti et al. 2016). Similarly, through the Bioconcentration Factor (BCF), phytoremediation efficiency of plants can be determined. It depicts the relationship between metal concentration in plants and soils (Melo et al. 2014). High BCF for Cd in the clayey loamy and silty loamy soil has been observed in *S. bicolor*.

Additionally, enhanced biomass concentration, high HM stress tolerance, 1.34 BCF value for Cd, 0.35 BCF value for Zn, and increased accumulation in shoots have been reported in *S. bicolor* (Epelde et al. 2009). Using black tea leaves residue, the phytoextraction capacity of two Sorghum species has been evaluated. Higher TF for different HMs was observed in *S. halepense* in comparison to *S. bicolor* (Ziarati et al. 2015). Similarly, high BCF, TF values, and 12–24 times increase in biomass content were observed in five Sorghum cultivars grown in toxic concentrations of Zn, Cd, and Pb (Yuan et al. 2019).

Several radionuclides such as uranium (U), cesium (Cs), and strontium (Sr) are present in the environment posing a serious threat to human health. Intriguingly, Sorghum has been reported to have BCF and TF values greater than 1 and accumulate huge amounts of Cs from the soil and water mediums with 86–92% of Cs accumulation in aerial parts of the plant without much impact on the biomass content (Wang et al. 2017c). Thus, *S. bicolor* exhibits metal accumulation potential and phytoremediation property due to its metal uptake and translocation ability. However, relatively little is known about the molecular basis of the accumulation potential of Sorghum. It is essential to understand the cellular/molecular mechanism of metal tolerance/accumulation and identify the rate-limiting phases for the uptake, translocation, and detoxification of metals in Sorghum.

3 Impact of Heavy Metals on the Morphological Characteristics of Sorghum

Heavy metals frequently induce morphological abnormalities in plants. The majority of morphological traits in Sorghum are associated with various economically important qualities, which help in selecting the high yielding Sorghum genotypes. Overall, the plant growth particularly related to root development significantly decreases due to HM toxicity (Borges et al. 2018; Hamim et al. 2018). At the tissue level, HM stress leads to modifications of the epidermis (Li et al. 2014), cortex, and vascular bundles (Daud et al. 2009). Differential response of plants has been observed upon exposure to various HMs. An inclusive dose-dependent decrease in shoot length, root length, fresh weight, and relative water content has been reported in *S. bicolor* seedlings under different concentrations of Cu stress (Roy et al. 2016b). Physiologically, the HM toxicity in plants decreases leaf size, lamina thickness, and size of stomatal aperture, which causes a reduction in transpiration and hence inhibits water uptake and translocation to shoot by plants (Rucińska-Sobkowiak 2016). Low concentration of Cd causes chlorosis of old leaves in Sorghum, without any significant alteration in shoot biomass and root dry weight. However, a high concentration of Cd causes a significant decrease in plant height, shoot biomass, root biomass, and rolling of young leaves. Ultrastructural damage and plasmolysis of the root cells also occur due to metal toxicity (Hamim et al. 2018). High Cd concentration leads to recognizable changes in the organelle structures like 2-fold, 2.5-fold, and 1.7-fold increase in the thickness of cell walls of vascular bundle cells, root xylem, and phloem, respectively. Also, an increase in starch grain in chloroplasts of vascular bundle sheath, the long-narrow chloroplast of mesophyll cells, and disarray in the arrangement of grana lamella in chloroplasts have been reported (Jia et al. 2016).

Interestingly, an increase in shoot length and reduction in plant height have been reported in two varieties of *S. bicolor*, Cowly and Negsi at 100 mg/kg and 400 mg/kg Cs, respectively. A significant decrease in dry weight of leaves, shoot, and root was observed in the two varieties at and above 400 mg/kg Cs concentration (Wang et al. 2017b). If the HM resistant microbe such as *Rhodococcus ruber* N7 is co-inoculated with *S. bicolor,* promising changes are observed under Cd stress. Withering of leaves and reduced root biomass have been prominently discerned in response to Cd stress (Muratova et al. 2015). Exposure to Cd and Zn leads to a significant reduction in the number of leaves, root growth, yellowing, and browning of leaves in *S. bicolor* (Soudek et al. 2014). A dose-dependent exposure of different concentrations of Cr in *S. bicolor* has shown a reduction in dry weight of root, shoot, and leaves with higher accumulation of Cr in root as compared to shoot (Malmir 2011). In sweet Sorghum and sudangrass, low Cd concentration showed a negligible effect on plant height, however, it gradually decreased with increasing Cd concentration (Da-lin et al. 2011). High concentrations of Cr^{3+} and Cr^{6+} induce inhibition of seed germination, root growth, and shoot biomass in wheat, oat, and Sorghum (López-Luna et al. 2009). A significant reduction in root and shoot biomass, as well as chlorotic symptoms, has been reported in *S. bicolor* in response to As stress. Though, *S. bicolor* is tolerant to low Cd concentration, the concentration above certain level causes necrosis in the tips of germinated seedlings, decrease in seed germination, root length, root hair density, panicle length, grain yield, grain dry weight, number of grains per panicle, and leaf senescence (Kuriakose and Prasad 2008). It is ascertained that HMs affect the concentration of essential elements in plants such as Potassium (K), Iron (Fe), Calcium (Ca), Magnesium (Mg), and Copper (Cu), which are required for various biological processes (Shaibur et al. 2008). The morphophysiological changes in Sorghum species under different HM stress have been summarized in Table 1.

4 Effect of Heavy Metals on the Physiological Aspects of Sorghum

Heavy metal phytotoxicity may result from alterations of numerous physiological processes caused at cellular/molecular level by inactivating enzymes, blocking functional groups of metabolically important molecules, displacing or substituting for essential elements, and disrupting membrane integrity (Rakhshaee et al. 2009; Kumar and Trivedi 2015b; Douchiche et al. 2010). Besides, HM toxicity causes inhibition of photosynthesis (Paunov et al. 2018), leaf senescence, alterations in photo-assimilate translocation, changes in water-relations, and so forth. The

Table 1 Morpho-physiological changes in Sorghum species under heavy metal stress

S. No.	Sorghum species	Heavy metals	Treatment Duration [days]	Conc.	Changes	References
1.	*Sorghum bicolor*	As	14	67 µM	Decrease in biomass content, cation uptake, and translocation; chlorosis	Shaibur et al. (2008)
		Cd	7; 100; 15; 28	3 mM; 15 mg/kg; 50 µM; 1,000 µM	Reduction in plant height, biomass content, growth, and seed germination; inhibited photosynthesis, electron transport, and chlorophyll synthesis	Kuriakose and Prasad (2008), Wang et al. (2017a), Soudek et al. (2014), Xue et al. (2018)
		Cr	7	500 mg/kg Cr^{3+}	Reduction in seed growth and germination	López-Luna et al. (2009)
			10	64 ppm	Decrease in chlorophyll content and plant biomass	Yilmaz et al. (2017)
		Cs	100	400 mg/kg	Inhibition of overall plant growth	Wang et al. (2016, 2017c)
		Cu	12	0.01–10 µM	High oxidative stress; reduction in photosynthesis	Székely et al. (2011)
		Ni	30	10 mg/L	Overall cessation of growth	Al Chami et al. (2015)
		Pb	7–30	100 mg/L	Decrease in chlorophyll and biomass content	Al Chami et al. (2015), Yilmaz et al. (2017)
		Sr	140	50–400 mg/kg	Increased plant height and biomass content	Wang et al. (2017b)
		U	60	20 mg/kg	Inhibition of protein and chlorophyll synthesis	Shi-Qi et al. (2018)
		Zn	28	1,000 µM	Reduction in chlorophyll content	Soudek et al. (2014)
2.	*Sorghum sudanense*	Cd	35	25 mg/kg	Decrease in plant height and chlorophyll content	Da-lin et al. (2011)
		Cu	12	0.01–10 µM	Oxidative stress and reduction in photosynthesis	Székely et al. (2011)

(continued)

Table 1 (continued)

S. No.	Sorghum species	Heavy metals	Treatment Duration [days]	Conc.	Changes	References
3.	*Sorghum hybrid sudangrass*	Cd	35	25 mg/kg	Decrease in plant height and chlorophyll content	Da-lin et al. (2011)
		U	60	20 mg/kg	Inhibition of protein and chlorophyll synthesis	Shi-Qi et al. (2018)

exposure to HMs significantly decreases the biomass and chlorophyll-a content in *Brassica juncea*. The effects of Pb, Cd, and Cr stress exert the minimal influence in mustard. On the contrary, the excess concentration of Ni hampers overall yield (Sheetal et al. 2016).

Heavy metal stress causes remarkable physiological changes in *S. bicolor* (Table 1). Low concentrations of U increase the chlorophyll content and soluble proteins in the shoot of Sorghum; however, the level decreases in response to high U concentration (Shi-Qi et al. 2018). A gradual decrease in chlorophyll content and resistance in electron transport in leaves has been observed in *S. bicolor* under increasing Cd concentrations (Xue et al. 2018). A dose-dependent decrease in photosynthesis, chlorophyll, and carotenoids content has been reported in *S. bicolor* (Jia et al. 2016). With increasing accumulation of Cu^{2+}, a notable decrease in concentrations of Zn^{2+}, Ca^{2+}, Fe^{2+} and Mn^{2+} occurs in *S. bicolor* seedlings exposed to different concentrations of Cu (Roy et al. 2016b). Additionally, an increase in *Chl a/b* ratio by conversion of *Chlb* to *Chla* and decrease in pigments, glucose, sucrose, soluble sugar, polysaccharide, and carotenoid content ensue in *S. bicolor* under Zn and Cd stress (Soudek et al. 2014; Aldesuquy et al. 2004). Notable reductions in the chlorophyll content in leaves of Sweet Sorghum, Sudangrass, and Sorghum hybrid sudangrass are observed in response to Cd (Da-lin et al. 2011). Moreover, analysis of maize and Sorghum roots under Cd stress discerned higher exudation of citrate and malate, respectively (Pinto et al. 2008). Overall, HMs when present above critical level compromise the crop yield and impose huge economic loss.

5 Biochemical Changes in Response to Heavy Metals in Sorghum

Plants exposed to high concentrations of HMs suffer from oxidative stress and disrupted metabolism at cellular levels. To counter these changes, plants produce certain stress-responsive proteins, amino acids, hormones, and antioxidant enzymes to overcome oxidative stress (Manara 2012; Kapoor et al. 2014). A gradual decrease in the activity of ascorbate peroxidase (APX), H_2O_2 concentration, and an increase in peroxidase (POX) activity have been related to Cr, Zn, and Cd heavy metals in plants (Sihag et al. 2016; Soudek et al. 2014). Malondialdehyde (MDA) content, lipid peroxidation, Glutathione S-transferase (GST) activity, and involvement of other antioxidant enzymes vary in *S. bicolor*, Sweet Sorghum, and Sorghum hybrid sudangrass under HM stress (Liu et al. 2010; Da-lin et al. 2011; Malmir 2011). Furthermore, GST and Glutathione Reductase (GR) hyperactivity help in mitigation of Cr^{6+} toxicity. GST assists in Cr^{6+} sequestration into the vacuoles and GR is suggested to help in the biosynthesis of reduced GSSG against Cr^{6+} toxicity (Malmir 2011). Acid phosphatase (ACP), proteases, and α-amylase activity significantly decrease in *S. bicolor* seedlings under Cd stress (Kuriakose and Prasad 2008).

Exposure of different concentrations of Cr^{3+} and Cr^{6+} on *S. bicolor* seedlings ascertains a significant increase in H_2O_2 concentrations and lipid peroxidation in all plant parts except leaves. Increase in Superoxide Dismutase (SOD), Dehydroascorbate Reductase (DHAR), and GR activity and no significant change in Catalase (CAT), APX, and Mono-dehydroascorbate Reductase (MDHAR) activity have been reported. A notable increase in amino acid, Glutathione (GSH), Dehydroascorbic Acid (DHA), and Glutathione disulfide (GSSG) levels has been ascertained under Cr^{3+} stress (Shanker and Pathmanabhan 2004). The above-mentioned modulations in the biochemical activities aid in better understanding the toxicity induced oxidative damage, imposition, and maintenance of ROS homeostasis under metal stress. Biochemical changes in *S. bicolor* under HM stress have been mentioned in Table 2.

6 Molecular Modulations Under Heavy Metal Stress in Sorghum

The plants have adapted sophisticated detoxification mechanisms, namely sequestration and compartmentalization of HMs in the vacuoles and synthesis of antioxidant enzymes to combat HM mediated stress (Arif et al. 2016; Dixit et al. 2015a; Kumar and Trivedi 2018; Rai et al. 2011; Shri et al. 2009; Verbruggen et al. 2009). In recent years, using molecular tools various genes and proteins involved in HM uptake and detoxification in plants have been identified and functionally characterized. In response to different stress conditions, plants initiate molecular changes, which help in adaptation. Heavy metals modulate the proteome and metabolome of plants and alter a functional translated portion of the genome (Ahsan et al. 2009; Dixit et al. 2015b; Villiers et al. 2011). Using next-generation tag sequencing, global gene transcripts have been identified in *Medicago truncatula* in response to Hg (Zhou et al. 2013). Similarly, genome-wide transcriptome profiling in rice roots on the exposure of Cd and Cr has been studied (Dubey et al. 2010; Lin et al. 2013). In the past decade, genetic engineering has escalated productivity and yield by conferring genetic traits to agro economically crops. Therefore, stress-sensitive plants can be tailored for tolerance by overexpression and/or ectopic expression of stress-

Table 2 Biochemical changes in *Sorghum bicolor* under different heavy metal stress

S. No.	Heavy metal	Conc.	Treatment duration (Days)	Enzyme	Tissues	Changes	Ref.
1.	Cd	100 μM	28	Glutathione-S-transferase	Shoots	Increase	Soudek et al. (2014)
		100 μM	28	Peroxidase	Roots		
		5,000 μM	28	Catalase	Roots and shoots		
		1,000 μM	28	Ascorbate peroxidase	Shoots		
2.	Cr	8 ppm	10	Ascorbate peroxidase	Roots and shoots	Increase	Shanker and Pathmanabhan (2004), Yilmaz et al. (2017), Sihag et al. (2016), Kumar and Joshi (2008)
		16 ppm	10	Catalase	Roots and shoots	Increase	
		50 μM Cr(IV)	10	Dihydroascorbate reductase	Roots	Increase	
		8 ppm	10	Glutathione-S-transferase	Roots and shoots	Increase	
		16 ppm	10	Glutathione reductase	Roots and shoots	Increase	
		50 μM Cr(IV)	10	Monodihydroascorbate reductase	Roots	Increase	
		8 ppm	10	Superoxide dismutase	Roots and shoots	Increase	
		4 ppm	90	Glutamate dehydrogenase	Roots and shoots	Decrease	
		2 ppm	90	Nitrate reductase	Roots and shoots	Increase	
		4 ppm	90	Glutamine synthetase	Roots and shoots	Decrease	
		1 ppm	90	Nitrite reductase	Roots and shoots	Decrease	
		4 mg/kg	90	Peroxidase	Shoots	Increase	
3.	Pb	4 ppm	10	Ascorbate peroxidase	Roots and shoots	Increase	Yilmaz et al. (2017)
		8 ppm	10	Catalase			
			10	Superoxide dismutase			
			10	Glutathione reductase			
4.	Zn	1,000 μM	28	Catalase	Roots and shoots	Increase	Soudek et al. (2014)
		1,000 μM	28	Ascorbate peroxidase	Shoots	Increase	
		5,000 μM	28	Glutathione-S-transferase	Shoots	Increase	
		100 μM	28	Peroxidase	Roots	Increase	

responsive genes from the resilient plants. Nonetheless, efficient genetic transformations in plants remain a challenge due to various factors. Molecular modifications need an in-depth understanding of the genes/proteins/metabolites and pathways that regulate stress tolerance in plants. Heavy metal is the most pervasive environmental problem and Sorghum shows tolerance towards an array of HMs. Therefore, comprehensive gene expression mining will attribute to the identification of candidate genes involved in providing metal tolerance. Few studies have underpinned the information related to the gene expression of Sorghum under different HMs (Table 3). Differential expression of 31 P_{1B}-ATPases (Heavy metal ATPases) genes has been identified in Sorghum, rice, and maize (Zhiguo et al. 2018). Under HM stress, the abscisic acid pathway with the help of ABI4 (Abscisic acid signaling cascade) up-regulates the expression of P_{1B}-ATPases. Increased Cd concentration induces the ZIP8 (Zinc-Iron Permease) pathway, which inhibits the HMA expression (Darabi et al. 2017). Under Cu stress, genes related to carbohydrate metabolism, stress defense, and protein translation get up-regulated, whereas genes related to energy metabolism, photosynthesis, plant growth, and development down-regulate in *S. bicolor* seedlings (Roy et al. 2016b, 2017). Exposure of *S. bicolor* seedlings to different concentrations of Cd has shown differential expression of genes (Abou-Elwafa et al. 2019; Roy et al. 2016a). Also, genetic diversity for Cd accumulation in Sorghum has been investigated. Through identification of quantitative trait loci (QTL) responsible for Cd accumulation in shoots, different landraces have been identified possessing a strong ability to absorb and translocate Cd (Tsuboi et al. 2017). The association mapping and expression analysis have identified 14 phytoremediation and HM tolerance QTLs in *S. bicolor* (Abou-Elwafa et al. 2019). Increased expression of Zinc Finger Proteins SbZFP17, SbLysMR1, SbPPR1 and SbZFP346. SbZFP17, and SbZFP346 has been observed in response to HM stress (Inui et al. 2015). SbPPR1, which is a homolog of the pentatricopeptide repeat containing proteins, has been identified to be associated with HMs stress tolerance. Up-regulation of SbHMA5b (under Cd, Cu, and Zn stress) involved in amino acid biosynthesis; SbHMA9/Cu-ion transmembrane transporter (under Cu stress) involved in secondary metabolite biosynthesis and SbHMA2/Cd-transporting ATPase (under Cd stress) involved in amino acid biosynthesis have been observed in *S. bicolor* under HM stress (Zhiguo et al. 2018; Darabi et al. 2017). Cinnamyl alcohol dehydrogenase, involved in cell wall reorganization, has been reported to be down-regulated in leaves and roots in response to Cd and Cu stress, respectively, in *S. bicolor* (Roy et al. 2016a, 2017). Also, up-regulation of glycolysis and carbohydrate metabolism-related proteins, glyceraldehyde-3-phosphate dehydrogenase, and Cytochrome P450 in leaves of *S. bicolor* under Cd and Cu stress has been manifested (Roy et al. 2016a, b). The modulation in the expression of different genes in *S. bicolor* due to HM stress has been summarized in Table 3. Thus, the exhaustive understanding of molecular mechanisms and detoxification pathways involved in providing tolerance to Sorghum is a prerequisite for developing plant varieties with genetic modifications that would impart tolerance towards different HMs.

Table 3 Differentially expressed genes in *Sorghum bicolor* under heavy metal stress

S. No.	Heavy metals	Conc.	Treatment duration (days)	Genes	Plant parts	Modulation	Function	Ref.
1.	Cd; Pb; Cu; Ni	10 mg/L	100	Receptor like protein kinase, Pentatricopeptide repeat containing protein 1, Zinc finger protein 346	Various tissues	Up-regulated	Heavy metal accumulation, phytoremediation	Abou-Elwafa et al. (2019)
2.	Cd; Pb; Cu; Ni	10 mg/L	100	Pentatricopeptide repeat containing protein, Zinc finger CCCH domain containing protein 18	Various tissues	Up-regulated	Heavy metal stress tolerance, accumulation	
3.	Cd; Pb; Cu; Ni	10 mg/L	100	Putative laccase-9, Zinc finger protein 8, Mitogen activated protein kinase kinase 5, Pyrophosphate-energized vacuolar membrane proton pump-like, Wall associated receptor kinase 2, Pentatricopeptide repeat containing protein 6	Various tissues	Up-regulated	Heavy metal stress tolerance	
4.	Cd; Pb; Cu; Ni	10 mg/L	100	7-deoxyloganetin-glucosyltransferase, 3-ketoacyl-CoA-synthase, Pyrophosphate-energized vacuolar membrane proton pump, Zinc finger protein 6, Pentatricopeptide repeat containing protein 7	Various tissues	Up-regulated	Heavy metal stress tolerance	
5.	Cd; Cu; Zn	10 mg/L	100	Heavy metal ATPase (HMA5b)	Various tissues	Up-regulated	Amino acid biosynthesis	
6.	Cu	10 mg/L	100	Heavy metal ATPase (HMA9)/Cu-ion transmembrane transporter	Various tissues	Up-regulated	Secondary metabolite biosynthesis	
7.	Cd	10 mg/L	100	Heavy metal ATPase (HMA2) Cd-transporting ATPase	Various tissues	Up-regulated	Amino acid biosynthesis	
8.	Cd	150 μmol/L	28	Cd-induced stress protein	Leaves	Up-regulated	Cd stress responsiveness	Metwali et al. (2013)

9.	Cd	100/ 150 µM	5	Cinnamyl alcohol dehydrogenase	Leaves	Down-regulated	Cell wall reorganization	Roy et al. (2016a)
		100/ 150 µM	5	Glyceraldehyde-3-phosphate dehydrogenase	Leaves	Down-regulated	Glycolysis and carbohydrate metabolism	
		100/ 150 µM	5	Cytochrome P450	Leaves	Up-regulated	Stress response	
10.	Cu	100/ 150 µM	5	Cinnamyl alcohol dehydrogenase	Roots	Down-regulated	Cell wall reorganization	Roy et al. (2017)
		100/ 150 µM	5	Cytochrome P450	Roots	Up-regulated	Stress response	
11.	Cu	100/ 150 µM	5	Glyceraldehyde-3-phosphate dehydrogenase	Leaves	Up-regulated	Glycolysis and carbohydrate metabolism	Roy et al. (2016b)

7 Future Prospects

The comprehensive "omics" based approaches could provide further insight into the molecular mechanisms underlying resilience in Sorghum against copious HMs. Extensive and focused research is necessary for understanding the genetic adaptability of Sorghum towards abiotic stresses. Thereafter, exploiting the molecular marker-assisted breeding techniques for developing resilient crops for phytoremediation should be the main target. The functional characterization of Sorghum HM tolerant loci shall help in the fine-tuning of the marker-assisted breeding approaches. Also, exploration of the detoxification mechanisms shall help to understand the stress tolerance upon HM uptake, transport, and accumulation. Several non-coding RNAs have been reported that play an important role in stress response in plants (Kumar et al. 2017; Sunkar et al. 2007; Basso et al. 2019). Therefore, studies on the expression and accumulation of non-coding RNAs in Sorghum shall elucidate the master regulators involved in providing tolerance towards different stress conditions. Also, Sorghum cultivation on a larger scale for phytoremediation of HMs and bio-ethanol production would solve the problem of environmental pollution due to fuel combustion. Besides, *S. bicolor* could be considered for its application as a possible rotation crop in HM contaminated soil for Rabi as well as Kharif crops. Therefore, the simple, monocot, HM tolerant *S. bicolor* holds a key to the future of food and environmental sustainability.

8 Conclusion

Heavy metal contamination is a serious anthropogenic issue disrupting the ecological food chain. It also affects plant growth and development exerting an economical pressure over arable land quality, crop productivity and yield. The ever increasing food demand and global hunger require an urgent need for soil and water remediation from an array of HMs. Sorghum, a hyperaccumulator of HMs is an essential biomass and energy crop grown predominantly in arid and semi-arid regions of the world with an ambit of adaptability to diverse agro-ecological conditions. It is a widely grown staple food and provides a cheap and gluten limited food source to the poverty-ridden regions for the world like Africa and some Asian countries. The Sorghum grain HM contamination has not been explored till date; however, the presence of various bacterial and fungal toxins have been reported commonly in grains, sorghum-based processed foods and byproducts that could be removed through various tweaking in food processing technology. Intriguingly, *S. bicolor* has the capability of accumulating high concentrations of almost all the HMs including various radioactive metals like Sr, Cs, and U (Wang et al. 2017b, c; Shi-Qi et al. 2018; Wang et al. 2016). Hence, Sorghum is an ideal model plant for studying plant adaptive responses to radioactive HM stress. Besides, *S. bicolor* possesses commendable phytoremediation capabilities comprising of phytoextraction, bioaccumulation, and phytostabilization. Therefore, it can be used for land reclamation by phytoremediation of radioactive metals contaminated soil.

Acknowledgments SK thankfully acknowledges the Science and Engineering Research Board (SERB), New Delhi, India for the Research Scientist Scheme. DM and BNM acknowledge the Institute of Engineering and Technology, APJ Abdul Kalam Technical University, India for the infrastructure and facilities.

On behalf of all authors, the corresponding author states that there is no conflict of interest.

References

Abou-Elwafa SF, Shehzad T (2018) Genetic identification and expression profiling of drought responsive genes in sorghum. Environ Exp Bot 155:12–20. https://doi.org/10.1016/j.envexpbot.2018.06.019

Abou-Elwafa SF, Amin AEEAZ, Shehzad T (2019) Genetic mapping and transcriptional profiling of phytoremediation and heavy metals responsive genes in sorghum. Ecotoxicol Environ Saf 173:366–372. https://doi.org/10.1016/j.ecoenv.2019.02.022

Adamczyk-Szabela D, Markiewicz J, Wolf WM (2015) Heavy metal uptake by herbs. IV. Influence of soil pH on the content of heavy metals in *Valeriana officinalis* L. Water Air Soil Pollut 226 (4). https://doi.org/10.1007/s11270-015-2360-3

Ahsan N, Renaut J, Komatsu S (2009) Recent developments in the application of proteomics to the analysis of plant responses to heavy metals. Proteomics 9(10):2602–2621. https://doi.org/10.1002/pmic.200800935

Al Chami Z, Amer N, Al Bitar L, Cavoski I (2015) Potential use of *Sorghum bicolor* and *Carthamus tinctorius* in phytoremediation of nickel, lead and zinc. Int J Environ Sci Technol 12 (12):3957–3970. https://doi.org/10.1007/s13762-015-0823-0

Aldesuquy HS, Haroun SA, Abo-Hamed SA, El-Saied AA (2004) Ameliorating effect of kinetin on pigments, photosynthetic characteristics, carbohydrate contents and productivity of cadmium treated *Sorghum bicolor* plants. Acta Bot Hungar 46:1–21. https://doi.org/10.1556/ABot.46.2004.1-2.1

Almodares A, Hadi MR, Kholdebarin B, Samedani B, Kharazian ZA (2014) The response of sweet sorghum cultivars to salt stress and accumulation of Na^+, Cl^- and K^+ ions in relation to salinity. J Environ Biol 35(4):733–739. http://europepmc.org/abstract/MED/25004761

Arif N, Yadav V, Singh S, Kushwaha BK, Singh S, Tripathi DK et al (2016) Assessment of antioxidant potential of plants in response to heavy metals. Plant Respons Xenobiotics:97–125. https://doi.org/10.1007/978-981-10-2860-1_5

Banerjee R, Goswami P, Pathak K, Mukherjee A (2016) Vetiver grass: an environment clean-up tool for heavy metal contaminated iron ore mine-soil. Ecol Eng 90:25–34. https://doi.org/10.1016/j.ecoleng.2016.01.027

Basso MF, Ferreira PCG, Kobayashi AK, Harmon FG, Nepomuceno AL, Molinari HBC, Grossi-de-Sa MF (2019) MicroRNAs and new biotechnological tools for its modulation and improving stress tolerance in plants. Plant Biotechnol J 17(8):1482–1500. https://doi.org/10.1111/pbi.13116

Bhatti SS, Kumar V, Singh N, Sambyal V, Singh J, Katnoria JK, Nagpal AK (2016) Physico-chemical properties and heavy metal contents of soils and kharif crops of Punjab, India. Procedia Environ Sci 35:801–808. https://doi.org/10.1016/j.proenv.2016.07.096

Borges KLR, Salvato F, Alcântara BK, Nalin RS, Piotto FÂ, Azevedo RA (2018) Temporal dynamic responses of roots in contrasting tomato genotypes to cadmium tolerance. Ecotoxicology 27(3):245–258. https://doi.org/10.1007/s10646-017-1889-x

Chakrabarty D, Trivedi PK, Misra P, Tiwari M, Shri M, Shukla D et al (2009) Comparative transcriptome analysis of arsenate and arsenite stresses in rice seedlings. Chemosphere 74(5):688–702. https://doi.org/10.1016/j.chemosphere.2008.09.082

Chibuike GU, Obiora SC (2014) Heavy metal polluted soils: effect on plants and bioremediation methods. Appl Environ Soil Sci 2014:1–12. https://doi.org/10.1155/2014/752708

Chopra R, Burow G, Burke JJ, Gladman N, Xin Z (2017) Genome-wide association analysis of seedling traits in diverse Sorghum germplasm under thermal stress. BMC Plant Biol 17(1):1–15. https://doi.org/10.1186/s12870-016-0966-2

Cong W, Miao Y, Xu L, Zhang Y, Yuan C, Wang J et al (2019) Transgenerational memory of gene expression changes induced by heavy metal stress in rice (*Oryza sativa* L.). BMC Plant Biol 19(1). https://doi.org/10.1186/s12870-019-1887-7

Da-lin L, Kai-qi H, Jing-jing M, Wei-wei Q, Xiu-ping W, Shu-pan Z (2011) Effects of cadmium on the growth and physiological characteristics of sorghum plants. Afr J Biotechnol 10(70):15770–15776. https://doi.org/10.5897/AJB11.848

Darabi SAS, Almodares A, Ebrahimi M (2017) In silico study shows arsenic induces P1B ATPase gene family as cation transporter by abscisic acid signaling pathway in seedling of *Sorghum bicolor*. Acta Physiol Plant 39(8). https://doi.org/10.1007/s11738-017-2472-z

Daud MK, Sun Y, Dawood M, Hayat Y, Variath MT, Wu YX et al (2009) Cadmium-induced functional and ultrastructural alterations in roots of two transgenic cotton cultivars. J Hazard Mater 161(1):463–473. https://doi.org/10.1016/j.jhazmat.2008.03.128

Demecsová L, Tamás L (2019) Reactive oxygen species, auxin and nitric oxide in metal-stressed roots: toxicity or defence. Biometals 32(5):717–744. https://doi.org/10.1007/s10534-019-00214-3

Diwan H, Ahmad A, Iqbal M (2012) Chromium-induced alterations in photosynthesis and associated attributes in Indian mustard. J Environ Biol 33:239–244. www.jeb.co.in

Dixit G, Singh AP, Kumar A, Dwivedi S, Deeba F, Kumar S et al (2015a) Sulfur alleviates arsenic toxicity by reducing its accumulation and modulating proteome, amino acids and thiol metabolism in rice leaves. Sci Rep 5:1–16. https://doi.org/10.1038/srep16205

Dixit G, Singh AP, Kumar A, Singh PK, Kumar S, Dwivedi S et al (2015b) Sulfur mediated reduction of arsenic toxicity involves efficient thiol metabolism and the antioxidant defense system in rice. J Hazard Mater 298:241–251. https://doi.org/10.1016/j.jhazmat.2015.06.008

Dixit G, Singh AP, Kumar A, Mishra S, Dwivedi S, Kumar S et al (2016) Reduced arsenic accumulation in rice (Oryza sativa L.) shoot involves sulfur mediated improved thiol metabolism, antioxidant system and altered arsenic transporters. Plant Physiol Biochem 99:86–96. https://doi.org/10.1016/j.plaphy.2015.11.005

Douchiche O, Driouich A, Morvan C (2010) Spatial regulation of cell-wall structure in response to heavy metal stress: cadmium-induced alteration of the methyl-esterification pattern of homogalacturonans. Ann Bot 105(3):481–491. https://doi.org/10.1093/aob/mcp306

Dubey S, Misra P, Dwivedi S, Chatterjee S, Bag SK, Mantri S et al (2010) Transcriptomic and metabolomic shifts in rice roots in response to Cr (VI) stress. BMC Genomics 11(1):648. https://doi.org/10.1186/1471-2164-11-648

Dutta S, Mitra M, Agarwal P, Mahapatra K, De S, Sett U, Roy S (2018) Oxidative and genotoxic damages in plants in response to heavy metal stress and maintenance of genome stability. Plant Signal Behav 13:1. https://doi.org/10.1080/15592324.2018.1460048

Epelde L, Mijangos I, Becerril JM, Garbisu C (2009) Soil microbial community as bioindicator of the recovery of soil functioning derived from metal phytoextraction with sorghum. Soil Biol Biochem 41(9):1788–1794. https://doi.org/10.1016/j.soilbio.2008.04.001

Feng J, Jia W, Lv S, Bao H, Miao F, Zhang X et al (2018) Comparative transcriptome combined with morpho-physiological analyses revealed key factors for differential cadmium accumulation in two contrasting sweet sorghum genotypes. Plant Biotechnol J 16(2):558–571. https://doi.org/10.1111/pbi.12795

Gajalakshmi S, Iswarya V, Ashwini R, Divya G, Mythili S, Sathiavelu A (2012) Evaluation of heavy metals in medicinal plants growing in Vellore District. Eur J Exp Biol 2(5):1457–1461

Galavi M, Jalali A, Ramroodi M, Mousavi SR, Galavi H (2010) Effects of treated municipal wastewater on soil chemical properties and heavy metal uptake by sorghum (Sorghum bicolor L.). J Agric Sci 2(3):235

Ghori NH, Ghori T, Hayat MQ, Imadi SR, Gul A, Altay V, Ozturk M (2019) Heavy metal stress and responses in plants. Int J Environ Sci Technol 16(3):1807–1828. https://doi.org/10.1007/s13762-019-02215-8

Hamim H, Miftahudin M, Setyaningsih L (2018) Cellular and ultrastructure alteration of plant roots in response to metal stress. In: Ratnadewi D, Hamim H (eds) Plant growth and regulation. IntechOpen. https://doi.org/10.5772/intechopen.79110

Inui H, Hirota M, Goto J, Yoshihara R, Kodama N, Matsui T et al (2015) Zinc finger protein genes from *Cucurbita pepo* are promising tools for conferring non-Cucurbitaceae plants with ability to accumulate persistent organic pollutants. Chemosphere 123:48–54. https://doi.org/10.1016/j.chemosphere.2014.11.068

Jia W, Lv S, Feng J, Li J, Li Y, Li S (2016) Morphophysiological characteristic analysis demonstrated the potential of sweet sorghum (*Sorghum bicolor* (L.) Moench) in the phytoremediation of cadmium-contaminated soils. Environ Sci Pollut Res 23(18):18823–18831. https://doi.org/10.1007/s11356-016-7083-5

Jia W, Miao F, Lv S, Feng J, Zhou S, Zhang X et al (2017) Identification for the capability of Cd-tolerance, accumulation and translocation of 96 sorghum genotypes. Ecotoxicol Environ Saf 145:391–397. https://doi.org/10.1016/j.ecoenv.2017.07.002

Kapoor D, Kaur S, Bhardwaj R (2014) Physiological and biochemical changes in *Brassica juncea* plants under Cd-induced stress. BioMed Res Int 2014. https://doi.org/10.1155/2014/726070

Kasim WA (2006) Changes induced by copper and cadmium stress in the anatomy and grain yield of Sorghum bicolor (L.) Moench. Int J Agric Biol 8(1):1–6. http://www.fspublishers.org

Kumar S, Joshi UN (2008) Nitrogen metabolism as affected by hexavalent chromium in sorghum (*Sorghum bicolor* L.). Environ Exp Bot 64:135–144. https://doi.org/10.1016/j.envexpbot.2008.02.005

Kumar S, Trivedi PK (2015a) Heavy metal stress signaling in plants. In: Plant metal interaction: emerging remediation techniques, vol 2. Elsevier, Amsterdam. https://doi.org/10.1016/B978-0-12-803158-2.00025-4

Kumar S, Trivedi PK (2015b) Transcriptome modulation in rice under abiotic stress. Plant Environ Interact:70–83. https://doi.org/10.1002/9781119081005.ch4

Kumar S, Trivedi PK (2018) Glutathione S-transferases: role in combating abiotic stresses including arsenic detoxification in plants. Front Plant Sci 9(June):1–9. https://doi.org/10.3389/fpls.2018.00751

Kumar S, Trivedi PK (2019) Genomics of arsenic stress response in plants. https://doi.org/10.1007/978-3-319-91956-0_10

Kumar S, Asif MH, Chakrabarty D, Tripathi RD, Trivedi PK (2011) Differential expression and alternative splicing of rice sulphate transporter family members regulate Sulphur status during plant growth, development and stress conditions. Funct Integr Genom 11:259–273. https://doi.org/10.1007/s10142-010-0207-y

Kumar S, Asif MH, Chakrabarty D, Tripathi RD, Dubey RS, Trivedi PK (2013a) Differential expression of Rice lambda class GST gene family members during plant growth, development, and in response to stress conditions. Plant Mol Biol Report 31:569–580. https://doi.org/10.1007/s11105-012-0524-5

Kumar S, Asif MH, Chakrabarty D, Tripathi RD, Dubey RS, Trivedi PK (2013b) Expression of a rice lambda class of glutathione S-transferase, OsGSTL2, in *Arabidopsis* provides tolerance to heavy metal and other abiotic stresses. J Hazard Mater 248-249:228–237. https://doi.org/10.1016/j.jhazmat.2013.01.004

Kumar S, Dubey RS, Tripathi RD, Chakrabarty D, Trivedi PK (2015) Omics and biotechnology of arsenic stress and detoxification in plants: current updates and prospective. Environ Int 74:221–230. https://doi.org/10.1016/j.envint.2014.10.019

Kumar S, Verma S, Trivedi PK (2017) Involvement of small RNAs in phosphorus and sulfur sensing, signaling and stress: current update. Front Plant Sci 8:1–12. https://doi.org/10.3389/fpls.2017.00285

Kumar S, Khare R, Trivedi PK (2019) Arsenic-responsive high-affinity rice sulphate transporter, OsSultr1;1, provides abiotic stress tolerance under limiting Sulphur condition. J Hazard Mater 373:753–762. https://doi.org/10.1016/j.jhazmat.2019.04.011

Kuriakose SV, Prasad MNV (2008) Cadmium stress affects seed germination and seedling growth in *Sorghum bicolor* (L.) Moench by changing the activities of hydrolyzing enzymes. Plant Growth Regul 54(2):143–156. https://doi.org/10.1007/s10725-007-9237-4

Li SW, Leng Y, Feng L, Zeng XY (2014) Involvement of abscisic acid in regulating antioxidative defense systems and IAA-oxidase activity and improving adventitious rooting in mung bean [*Vigna radiata* (L.) Wilczek] seedlings under cadmium stress. Environ Sci Pollut Res 21 (1):525–537. https://doi.org/10.1007/s11356-013-1942-0

Lin CY, Trinh NN, Fu SF, Hsiung YC, Chia LC, Lin CW, Huang HJ (2013) Comparison of early transcriptome responses to copper and cadmium in rice roots. Plant Mol Biol 81(4–5):507–522. https://doi.org/10.1007/s11103-013-0020-9

Liu DL, Zhang SP, Chen Z, Qiu WW (2010) Soil cadmium regulates antioxidases in Sorghum. Agric Sci China 9(10):1475–1480. https://doi.org/10.1016/S1671-2927(09)60240-6

López-Luna J, González-Chávez MC, Esparza-García FJ, Rodríguez-Vázquez R (2009) Toxicity assessment of soil amended with tannery sludge, trivalent chromium and hexavalent chromium, using wheat, oat and sorghum plants. J Hazard Mater 163(2–3):829–834. https://doi.org/10.1016/j.jhazmat.2008.07.034

Lorestani B, Yousefi N, Cheraghi M, Farmany A (2013) Phytoextraction and phytostabilization potential of plants grown in the vicinity of heavy metal-contaminated soils: a case study at an industrial town site. Environ Monit Assess 185(12):10217–10223. https://doi.org/10.1007/s10661-013-3326-9

Malmir HA (2011) Comparison of antioxidant enzyme activities in leaves stems and roots of Sorghum (*Sorghum bicolor* L.) exposed to chromium (VI). African J Plant Sci 5(8):436–444. http://www.academicjournals.org/ajps

Manara A (2012) Plant responses to heavy metal toxicity. https://doi.org/10.1007/978-94-007-4441-7_2

Marchiol L, Fellet G, Perosa D, Zerbi G (2007) Removal of trace metals by *Sorghum bicolor* and *Helianthus annuus* in a site polluted by industrial wastes: a field experience. Plant Physiol Biochem 45(5):379–387. https://doi.org/10.1016/j.plaphy.2007.03.018

Masindi V, Muedi KL (2018) Environmental contamination by heavy metals. In: Saleh HE-DM, Aglan RF (eds) Heavy metals. https://doi.org/10.5772/intechopen.76082

Melo LCA, Silva EBD, Alleoni LRF (2014) Transfer of cadmium and barium from soil to crops grown in tropical soils. Rev Bras Ciênc Solo 38(6):1939–1949. https://doi.org/10.1590/S0100-06832014000600028

Metwali EMR, Gowayed SMH, Al-Maghrabi OA, Mosleh YY (2013) Evaluation of toxic effect of copper and cadmium on growth, physiological traits and protein profile of wheat (Triticum aestivium L.), maize (Zea mays L.) and shorghum (Sorghum bicolor L.). World Appl Sci J 21:301–314.

Migeon A, Blaudez D, Wilkins O, Montanini B, Campbell MM, Richaud P et al (2010) Genome-wide analysis of plant metal transporters, with an emphasis on poplar. Cell Mol Life Sci 67 (22):3763–3784. https://doi.org/10.1007/s00018-010-0445-0

Mittler R (2002) Oxidative stress, antioxidants and stress tolerance. Trends Plant Sci 7(9):405–410. https://doi.org/10.1016/S1360-1385(02)02312-9

Muratova A, Lyubun Y, German K, Turkovskaya O (2015) Effect of cadmium stress and inoculation with a heavy-metal-resistant bacterium on the growth and enzyme activity of *Sorghum bicolor*. Environ Sci Pollut Res 22(20):16098–16109. https://doi.org/10.1007/s11356-015-4798-7

Nowack B, Schwyzer I, Schulin R (2008) Uptake of Zn and Fe by wheat (Triticum aestivum var. Greina) and transfer to the grains in the presence of chelating agents (ethylenediaminedisuccinic acid and ethylenediaminetetraacetic acid). J Agric Food Chem 56(12):4643–4649. https://doi.org/10.1021/jf800041b

Oh K, Cao T, Cheng H, Liang X, Hu X, Yan L, Yonemochi S, Takahi S (2015) Phytoremediation potential of Sorghum as a biofuel crop and the enhancement effects with microbe inoculation in heavy metal contaminated soil. J Biosci Med 3:9–14. https://doi.org/10.4236/jbm.2015.36002

Padmapriya S, Murugan N, Ragavendran C, Thangabalu R, Natarajan D (2016) Phytoremediation potential of some agricultural plants on heavy metal contaminated mine waste soils, Salem District, Tamil Nadu. Int J Phytoremediation 18(3):288–294. https://doi.org/10.1080/15226514.2015.1085832

Paterson AH, Bowers JE, Bruggmann R, Dubchak I, Grimwood J, Gundlach H et al (2009) The *Sorghum bicolor* genome and the diversification of grasses. Nature 457(7229):551–556. https://doi.org/10.1038/nature07723

Paunov M, Koleva L, Vassilev A, Vangronsveld J, Goltsev V (2018) Effects of different metals on photosynthesis: cadmium and zinc affect chlorophyll fluorescence in durum wheat. Int J Mol Sci 19(3). https://doi.org/10.3390/ijms19030787

Phuong N, Afolayan G, Stützel H, Uptmoor R, El-Soda M (2019) Unraveling the genetic complexity underlying sorghum response to water availability. PLoS One 14(4):1–15. https://doi.org/10.1371/journal.pone.0215515

Pinto AP, Mota AM, de Varennes SA, Pinto FC (2004) Influence of organic matter on the uptake of cadmium, zinc, copper and iron by sorghum plants. Sci Total Environ 326:239–247. https://doi.org/10.1016/j.scitotenv.2004.01.004

Pinto AP, Vilar MT, Pinto FC, Mota AM (2005) Organic matter influence in cadmium uptake by Sorghum. J Plant Nutr 27(12):2175–2188. https://doi.org/10.1081/PLN-200034681

Pinto AP, Simões I, Mota AM (2008) Cadmium impact on root exudates of sorghum and maize plants: a speciation study. J Plant Nutr 31(10):1746–1755. https://doi.org/10.1080/01904160802324829

Rai A, Tripathi P, Dwivedi S, Dubey S, Shri M, Kumar S et al (2011) Arsenic tolerances in rice (*Oryza sativa*) have a predominant role in transcriptional regulation of a set of genes including Sulphur assimilation pathway and antioxidant system. Chemosphere 82:986–995. https://doi.org/10.1016/j.chemosphere.2010.10.070

Rai PK, Lee SS, Zhang M, Tsang YF, Kim KH (2019) Heavy metals in food crops: health risks, fate, mechanisms, and management. Environ Int:365–385. https://doi.org/10.1016/j.envint.2019.01.067

Rakhshaee R, Giahi M, Pourahmad A (2009) Studying effect of cell wall's carboxyl-carboxylate ratio change of Lemna minor to remove heavy metals from aqueous solution. J Hazard Mater 163(1):165–173. https://doi.org/10.1016/j.jhazmat.2008.06.074

Rekik I, Chaabane Z, Missaoui A, Bouket AC, Luptakova L, Elleuch A, Belbahri L (2017) Effects of untreated and treated wastewater at the morphological, physiological and biochemical levels on seed germination and development of sorghum (*Sorghum bicolor* (L.) Moench), alfalfa (*Medicago sativa* L.) and fescue (*Festuca arundinacea* Schreb.). J Hazard Mater 326:165–176. https://doi.org/10.1016/j.jhazmat.2016.12.033

Rooney WL (2014) 7 Sorghum. In: Cellulosic energy cropping systems, pp 109–129. http://faostat.fao.org/

Roy SK, Cho SW, Kwon SJ, Kamal AHM, Kim SW, Oh MW et al (2016a) Morpho-physiological and proteome level responses to cadmium stress in sorghum. PLoS One 11(2). https://doi.org/10.1371/journal.pone.0150431

Roy SK, Kwon SJ, Cho SW, Kamal AHM, Kim SW, Sarker K et al (2016b) Leaf proteome characterization in the context of physiological and morphological changes in response to copper stress in sorghum. Biometals 29(3):495–513. https://doi.org/10.1007/s10534-016-9932-6

Roy SK, Cho SW, Kwon SJ, Kamal AHM, Lee DG, Sarker K et al (2017) Proteome characterization of copper stress responses in the roots of sorghum. Biometals 30(5):765–785. https://doi.org/10.1007/s10534-017-0045-7

Rucińska-Sobkowiak R (2016) Water relations in plants subjected to heavy metal stresses. Acta Physiol Plant 38. https://doi.org/10.1007/s11738-016-2277-5

Seregin IV, Shpigun LK, Ivanov VB (2004) Distribution and toxic effects of cadmium and lead on maize roots. Russ J Plant Physiol 51(4):525–533. https://doi.org/10.1023/B:RUPP. 0000035747.42399.84

Shahid M, Austruy A, Echevarria G, Arshad M, Sanaullah M, Aslam M et al (2014) EDTA-enhanced phytoremediation of heavy metals: a review. Soil Sediment Contam Int J 23 (4):389–416. https://doi.org/10.1080/15320383.2014.831029

Shaibur MR, Kitajima N, Sugawara R, Kondo T, Alam S, Imamul Huq SM, Kawai S (2008) Critical toxicity level of arsenic and elemental composition of arsenic-induced chlorosis in hydroponic sorghum. Water Air Soil Pollut 191(1–4):279–292. https://doi.org/10.1007/s11270-008-9624-0

Shanker AK, Pathmanabhan G (2004) Speciation dependent antioxidative response in roots and leaves of sorghum (*Sorghum bicolor* (L.) Moench cv CO 27) under Cr (III) and Cr (VI) stress. Plant Soil 265

Sheetal KR, Singh SD, Anand A, Prasad S (2016) Heavy metal accumulation and effects on growth, biomass and physiological processes in mustard. Indian J Plant Physiol 21(2):219–223. https://doi.org/10.1007/s40502-016-0221-8

Shi-Qi Z, Li-Xing W, Qing H, Peng Z Li-Shan R (2018) Study on growth and accumulation characteristics of *Sorghum bicolor* × *S.Sudanense* on uranium. In: 2018 7th international conference on energy, environment and sustainable development (ICEESD 2018). Atlantis Press

Shri M, Kumar S, Chakrabarty D, Trivedi PK, Mallick S, Misra P et al (2009) Effect of arsenic on growth, oxidative stress, and antioxidant system in rice seedlings. Ecotoxicol Environ Saf 72 (4):1102–1110. https://doi.org/10.1016/j.ecoenv.2008.09.022

Shukla T, Khare R, Kumar S, Lakhwani D, Sharma D, Asif MH, Trivedi PK (2018) Differential transcriptome modulation leads to variation in arsenic stress response in *Arabidopsis thaliana* accessions. J Hazard Mater 351(2010):1–10. https://doi.org/10.1016/j.jhazmat.2018.02.031

Sihag S, Wadhwa N, Joshi UN (2016) Chromium toxicity affects antioxidant enzyme activity in *Sorghum bicolor* (L.). Forage Res 42

Singh R, Singh DP, Kumar N, Bhargava SK, Barman SC (2010) Accumulation and translocation of heavy metals in soil and plants from fly ash contaminated area. J Environ Biol 31(4):421–430

Soudek P, Nejedlý J, Pariči L, Petrová Š, Vaněk T (2013) The sorghum plants utilization for accumulation of heavy metals. Int Proc Chem Biol Environ Eng (IPCBEE) 54:6–11

Soudek P, Petrová Š, Vaňková R, Song J, Vaněk T (2014) Accumulation of heavy metals using *Sorghum sp.* Chemosphere 104:15–24. https://doi.org/10.1016/j.chemosphere.2013.09.079

Sunkar R, Chinnusamy V, Zhu J, Zhu JK (2007) Small RNAs as big players in plant abiotic stress responses and nutrient deprivation. Trends Plant Sci 12(7):301–309. https://doi.org/10.1016/j.tplants.2007.05.001

Székely Á, Poór P, Bagi I, Csiszár J, Gémes K, Horváth F, Tari I (2011) Effect of EDTA on the growth and copper accumulation of sweet sorghum and sudangrass seedlings. Acta Biol Szegediensis 55. http://www.sci.u-szeged.hu/ABSARTICLE

Tang J, Zhang J, Ren L, Zhou Y, Gao J, Luo L et al (2019) Diagnosis of soil contamination using microbiological indices: a review on heavy metal pollution. J Environ Manag 242 (April):121–130. https://doi.org/10.1016/j.jenvman.2019.04.061

Tsuboi K, Shehzad T, Yoneda J, Uraguchi S, Ito Y, Shinsei L et al (2017) Genetic analysis of cadmium accumulation in shoots of sorghum landraces. Crop Sci 57(1):22–31. https://doi.org/10.2135/cropsci2016.01.0069

Verbruggen N, Hermans C, Schat H (2009) Mechanisms to cope with arsenic or cadmium excess in plants. Curr Opin Plant Biol 12(3):364–372. https://doi.org/10.1016/j.pbi.2009.05.001

Villiers F, Ducruix C, Hugouvieux V, Jarno N, Ezan E, Garin J et al (2011) Investigating the plant response to cadmium exposure by proteomic and metabolomic approaches. Proteomics 11 (9):1650–1663. https://doi.org/10.1002/pmic.201000645

Wang X, Chen C, Wang J (2016) Bioremediation of cesium-contaminated soil by *Sorghum bicolor* and soil microbial community analysis. Geomicrobiol J 33(3–4):216–221. https://doi.org/10.1080/01490451.2015.1067655

Wang X, Chen C, Wang J (2017a) Cadmium phytoextraction from loam soil in tropical southern China by *Sorghum bicolor*. Int J Phytoremediation 19(6):572–578

Wang X, Chen C, Wang J (2017b) Phytoremediation of strontium contaminated soil by *Sorghum bicolor* (L.) Moench and soil microbial community-level physiological profiles (CLPPs). Environ Sci Pollut Res 24(8):7668–7678. https://doi.org/10.1007/s11356-017-8432-8

Wang X, Chen C, Wang J, Lou K (2017c) Cs phytoremediation by *Sorghum bicolor* cultivated in soil and in hydroponic system. Int J Phytoremediation 19(4):402–412

Xie Y, Ye S, Wang Y, Xu L, Zhu X, Yang J et al (2015) Transcriptome-based gene profiling provides novel insights into the characteristics of radish root response to Cr stress with next-generation sequencing. Front Plant Sci 6(March):1–13. https://doi.org/10.3389/fpls.2015.00202

Xue ZC, Li JH, Li DS, Li SZ, Jiang CD, Liu LA et al (2018) Bioaccumulation and photosynthetic activity response of sweet sorghum seedling (*Sorghum bicolor* L. Moench) to cadmium stress. Photosynthetica 56(4):1422–1428

Yaashikaa PR, Kumar PS, Saravanan A (2019) Modeling and Cr (VI) ion uptake kinetics of *Sorghum bicolor* plant assisted by plant growth–promoting *Pannonibacter phragmetitus*: an ecofriendly approach. Environ Sci Pollut Res 27:27307. https://doi.org/10.1007/s11356-019-05764-0

Yilmaz SH, Kaplan M, Temizgul R, Yilmaz S (2017) Antioxidant enzyme response of sorghum plant upon exposure to aluminum, chromium and Lead heavy metals. Turk J Biochem 42 (4):503–512. https://doi.org/10.1515/tjb-2016-0112

Yuan X, Xiong T, Yao S, Liu C, Yin Y, Li H, Li N (2019) A real filed phytoremediation of multi-metals contaminated soils by selected hybrid sweet sorghum with high biomass and high accumulation ability. Chemosphere 237:124536. https://doi.org/10.1016/j.chemosphere.2019.124536

Zheng LY, Guo XS, He B, Sun LJ, Peng Y, Dong SS et al (2011) Genome-wide patterns of genetic variation in sweet and grain sorghum (*Sorghum bicolor*). Genome Biol 12(11):R114. https://doi.org/10.1186/gb-2011-12-11-r114

Zhiguo E, Tingting L, Chen C, Lei W (2018) Genome-wide survey and expression analysis of P 1B -ATPases in Rice, maize and Sorghum. Rice Sci 25(4):208–217. https://doi.org/10.1016/j.rsci.2018.06.004

Zhou ZS, Yang SN, Li H, Zhu CC, Liu ZP, Yang ZM (2013) Molecular dissection of mercury-responsive transcriptome and sense/antisense genes in *Medicago truncatula*. J Hazard Mater 252–253:123–131. https://doi.org/10.1016/j.jhazmat.2013.02.011

Zhuang P, Shu W, Li Z, Liao B, Li J, Shao J (2009) Removal of metals by sorghum plants from contaminated land. J Environ Sci 21(10):1432–1437. https://doi.org/10.1016/S1001-0742(08)62436-5

Ziarati P, Ziarati NN, Nazeri S, Saber-Germi M (2015) Phytoextraction of heavy metals by two Sorghum spices in treated soil "using black tea residue for cleaning-Uo the contaminated soil". Orient J Chem 31(1):317–326. https://doi.org/10.13005/ojc/310136

Water and Soil Pollution: Ecological Environmental Study Methodologies Useful for Public Health Projects. A Literature Review

Roberto Lillini, Andrea Tittarelli, Martina Bertoldi, David Ritchie, Alexander Katalinic, Ron Pritzkuleit, Guy Launoy, Ludivine Launay, Elodie Guillaume, Tina Žagar, Carlo Modonesi, Elisabetta Meneghini, Camilla Amati, Francesca Di Salvo, Paolo Contiero, Alessandro Borgini, and Paolo Baili

Contents

1	Introduction	181
2	Methods	182
3	Results	201
	3.1 Type 1: Regression with Data by Geographical Area	201
	3.2 Type 2: Regression Models at Individual Level	203
	3.3 Type 3: Exposure Intensity Threshold Values for Evaluating Health Outcome	204
	3.4 Type 4: Distance between Pollutant Source and Health Outcome Clusters	205
4	Discussion	206
5	Conclusion	210
	References	211

WASABY Working Group have been listed in the acknowledgments.

Roberto Lillini, Andrea Tittarelli, Alessandro Borgini, and Paolo Baili contributed equally to this work.

R. Lillini (✉) · E. Meneghini · C. Amati · P. Baili
Analytical Epidemiology and Health Impact Unit, Fondazione IRCCS "Istituto Nazionale dei Tumori", Milan, Italy
e-mail: roberto.lillini@istitutotumori.mi.it; elisabetta.meneghini@istitutotumori.mi.it; lifetable@istitutotumori.mi.it

A. Tittarelli
Cancer Registry Unit, Fondazione IRCCS "Istituto Nazionale dei Tumori", Milan, Italy
e-mail: andrea.tittarelli@istitutotumori.mi.it

M. Bertoldi · P. Contiero
Environmental Epidemiology Unit, Fondazione IRCCS "Istituto Nazionale dei Tumori", Milan, Italy
e-mail: martina.bertoldi@istitutotumori.mi.it; paolo.contiero@istitutotumori.mi.it

D. Ritchie
Association Européenne des Ligues contre le Cancer, Bruxelles, Belgium
e-mail: david@europeancancerleagues.org

© The Author(s), under exclusive license to Springer Nature Switzerland AG 2020
P. de Voogt (ed.), *Reviews of Environmental Contamination and Toxicology Volume 256*, Reviews of Environmental Contamination and Toxicology 256,
https://doi.org/10.1007/398_2020_58

Abstract Health risks at population level may be investigated with different types of environmental studies depending on access to data and funds. Options include ecological studies, case–control studies with individual interviews and human sample analysis, risk assessment or cohort studies. Most public health projects use data and methodologies already available due to the cost of ad-hoc data collection. The aim of the article is to perform a literature review of environmental exposure and health outcomes with main focus on methodologies for assessing an association between water and/or soil pollutants and cancer. A systematic literature search was performed in May 2019 using PubMed. Articles were assessed by four independent reviewers. Forty articles were identified and divided into four groups, according to the data and methods they used, i.e.: (1) regression models with data by geographical area; (2) regression models with data at individual level; (3) exposure intensity threshold values for evaluating health outcome trends; (4) analyses of distance between source of pollutant and health outcome clusters. The issue of exposure

A. Katalinic · R. Pritzkuleit
Institute for Cancer Epidemiology at the University Lübeck, Lübeck, Germany
e-mail: Alexander.Katalinic@uksh.de; Ron.Pritzkuleit@uksh.de

G. Launoy
Normandie Univ, UNICAEN, INSERM, ANTICIPE, Caen, France

Pôle recherche – Centre Hospitalier Universitaire, Caen, France
e-mail: guy.launoy@unicaen.fr

L. Launay
Normandie Univ, UNICAEN, INSERM, ANTICIPE, Caen, France

Centre François Baclesse, Caen, France
e-mail: ludivine.launay@inserm.fr

E. Guillaume
Normandie Univ, UNICAEN, INSERM, ANTICIPE, Caen, France
e-mail: elodie.guillaume@unicaen.fr

T. Žagar
Institute of Oncology Ljubljana, Ljubljana, Slovenia
e-mail: TZagar@onko-i.si

C. Modonesi
Cancer Registry Unit, Fondazione IRCCS "Istituto Nazionale dei Tumori", Milan, Italy

International Society of Doctors for the Environment (ISDE), Arezzo, Italy
e-mail: carlo.modonesi@istitutotumori.mi.it

F. Di Salvo
Pancreas Translational and Clinical Research Center, Ospedale IRCCS "San Raffaele", Milan, Italy
e-mail: disalvo.francesca@hsr.it

A. Borgini
Environmental Epidemiology Unit, Fondazione IRCCS "Istituto Nazionale dei Tumori", Milan, Italy

International Society of Doctors for the Environment (ISDE), Arezzo, Italy
e-mail: alessandro.borgini@istitutotumori.mi.it

assessment has been investigated for over 40 years and the most important innovations regard technologies developed to measure pollutants, statistical methodologies to assess exposure, and software development. Thanks to these changes, it has been possible to develop and apply geo-coding and statistical methods to reduce the ecological bias when considering the relationship between humans, geographic areas, pollutants, and health outcomes. The results of the present review may contribute to optimize the use of public health resources.

Keywords Health outcomes · Public health · Soil pollution · Spatial analysis · Statistical methods · Water pollution

1 Introduction

The effects of the industrialization of many economic activities during the last two centuries have become an important issue for environmental and human health. As recognized by the World Health Organization (WHO), environmental integrity is a major determinant of health and actions to protect human and animal populations from diseases should be a primary objective of a global health agenda, particularly in the case of degenerative pathologies. Based on the above principles, the European health policy framework "Health 2020", supported by WHO's Regional Office for Europe, aims to improve the health and well-being of European citizens by reducing the weight of all disease determinants (World Health Organization Website 2019). Among actions promoted by Health 2020, management and remediation policies for preserving the resilient functions of ecosystems and environmental matrices are crucial.

A challenging issue lies in the considerable health risk resulting from the exposure to toxic chemicals and other stressors of industrial origin, as documented by a number of studies developed also in Europe (Hänninen et al. 2014). This kind of investigation can prove problematic, as people are exposed to hundreds of toxicants that come from both anthropogenic and natural sources: their physical and chemical interactions determine an extremely complex picture of phenomena that must necessarily to be taken into account. Contaminants move across environmental matrices and often accumulate in the organisms therein. The assessment of potential health effects due to exposure to all factors is a demanding task, often one too complex to be performed. Some chemicals are widespread on a global scale, while others accumulate around industrial or other specific sites; in this case, their concentration significantly exceeds that of background values. This results in considerable disparities in the level of exposure of human populations (Stewart and Wild 2014) and increases the obstacles when trying to explore the relationship between environmental pollution and health outcomes. However, an appropriate epidemiologic approach can contribute to clarify causes of disease, factors conferring susceptibility, and actual levels of exposure at which health effects occur (Deener et al. 2018).

Health risks at population level may be investigated with different types of environmental studies depending on access to data and funds. Options include ecological studies, case–control studies with individual interviews and human sample analysis, risk assessment or cohort studies. (Baker et al. 1999).

In 2016, the Health and Food Safety Directorate General (DG SANTE) of the European Commission launched, under the 3rd Health Programme, a call for project proposals aiming to identify geographical regions presenting higher breast cancer rates within the European Union, and to investigate the statistical correlation between water and soil polluting agents and high cancer rates (European Commission 2016). The *WASABY – Water And Soil contamination and Awareness on Breast cancer risk in Young women* project was established with the following objectives: (1) mapping breast cancer risk to identify areas at higher risk using specific geographic information systems; (2) reviewing scientific literature on the relationship between water and soil pollutants and breast cancer risk, and on possible methods for a pilot ecological environmental study (WASABY Website 2019). We defined the above objectives in consideration of the scopes of a DG SANTE call (i.e., excluding analytic studies such as cohort or case–control studies which could aim to evaluate cause-and-effect relationship) and of the call budget (European Commission 2016).

As most public health projects, WASABY focuses its activities using available data and methodologies (i.e., incidence data from cancer registries and databases of environmental agencies, spatial mapping methods, and ecological regression methods used in environmental studies).

In the present article, we summarize a PubMed (National Center for Biotechnology Information 2019) literature review of methodologies applied across the world to study the correlation between water and soil pollutants (e.g., arsenic in water, topsoil metals, etc.) and a given health outcome (e.g., cancer incidence, acute gastrointestinal infection hospital admissions, etc.) using available data. The review included all methodologies regardless of the aim of the environmental studies. For these reasons, we expected all or most of the articles considered in the review to be about cross-sectional studies. Focus of the review was to identify and describe materials, methods, and software programs. Therefore, the review does not present specific results.

2 Methods

In May 2019, we carried out a systematic literature search on articles describing ecological environmental methods using PubMed (National Center for Biotechnology Information 2019).

After a few tests to assess the most appropriate search terms to be used, we applied the following sequence of terms related with logical operators: "*(spatial analysis OR geographic analysis OR GIS) AND (water pollution OR water pollutants OR soil pollutants) AND (cancer registry OR population-based OR estimate OR estimating OR cancer incidence OR cancer mortality).*"

As a second step, we defined exclusion criteria, as follows: (a) articles on air pollutants not included in the project aims; (b) articles without real health outcome data such as risk assessment studies; (c) articles with ad-hoc data collection such as interviews or blood tests; (d) articles without spatial analysis; (e) articles not published in English.

The article revision process followed three phases. In *Phase 1*, three reviewers independently examined the abstracts of the articles identified by the PubMed search, so as to identify those potentially pertaining to our project aims. Articles would be considered eligible for Phase 2 if they were cleared by at least one of the reviewers. In *Phase 2*, four reviewers independently read the complete articles identified in *Phase 1*. In *Phase 3*, the reviewers met to address any divergence over *Phase 2* revisions.

At this stage, we then described the articles according to the following topics: country (or region) where the study was conducted; health outcome (dependent variable); environmental factors under analysis; socio-economic variables considered; smallest area unit considered for dependent variable; smallest area unit considered for environmental factor(s); final smallest area unit considered in the analysis; methods used; software used. Finally, we classified all selected papers into four subgroups, according to the methodology used and/or the data considered (a summary of characteristics is provided in Tables 1, 2, 3, 4, and 5 show articles by group).

Table 1 Synthesis of type of analysis to be performed for feasibility studies between water and soil pollutants and health outcomes, according to available data

	Environmental factor data	Health outcome data	Analysis	Number of articles
Type 1	Data by geographical areas	Data by geographical areas	Regression models using data by geographical areas	20
Type 2	Data at individual level	Data at individual level	Regression models using data at individual level	4
Type 3	Data by geographical areas	Data by geographical areas	Threshold values for exposure intensity are computed, in order to define cut-off points for evaluating trends in the health outcome variable influenced by the environmental factor	9
Type 4	Environmental pollution geographic clusters obtained by considering environmental factors and their potential emission sources	Clusters of areas or people generated by the analysis of the considered health outcomes	The two different kinds of clusters were identified separately. Comparisons between health outcomes geographic clusters and environmental pollution geographic clusters by considering the distance between them	7

Table 2 Articles classified as Type 1 and main characteristics

Reference PMID	Title	Country/region	Dependent variable	Environmental factors	Socio-economic variables	Smallest area unit (dep. variable)	Smallest area unit (envir. factor)	Final smallest area unit considered	Methods	Software
Aballay et al. (2012); PMID: 22017596	Cancer incidence and pattern of arsenic concentration in drinking water wells in Cordoba, Argentina.	Cordoba province (Argentina)	5 cancer sites incidence	Arsenic in water	Gender, age, urban/rural residence	Districts	Sampling points in districts	Districts	Generalized linear latent and mixed model (GLLAMM). Likelihood ratio tests (LRT) were performed using the equivalent Poisson regression model for the random intercept model. Statistical significance at $p < 0.01$	STATA 10 with xtmepoisson command
Armijo and Coulson (1975); PMID: 23682416	Epidemiology of stomach cancer in Chile – The role of nitrogen fertilizers.	Chile	Mortality by stomach cancer	Nitrates in drinking water and nitrogen fertilizers	Infant mortality rates, housing ratings	Province	Province	Province	Bivariate correlation. Statistical significance at $p < 0.05$	Not declared
Bulka et al. (2016); PMID: 27136670	Arsenic in drinking water and prostate cancer in Illinois counties: An ecologic study.	Illinois state Cancer registry (USA)	Prostate cancer incidence	Arsenic (in drinking water)	Percent of individuals in the county living under the federal poverty level	County	County	County	Poisson regression model with robust standard errors. The model residuals were tested for spatial autocorrelation by calculating a global Moran's I statistic. Statistical significance at $p < 0.05$, $p < 0.01$	SAS 9.4

Study	Title	Location	Outcome	Exposure	Sex/Race restriction	Exposure unit	Outcome unit	Analysis unit	Methods	Software
Chiang et al. (2010); PMID: 21139868	Spatiotemporal trends in oral cancer mortality and potential risks associated with heavy metal content in Taiwan soil.	Taiwan	Oral cancers age-standardized mortality rates	8 heavy metals (As, Cd, Cr, Cu, Hg, Ni, Pb, Zn) in soil	No	Townships	Townships	Townships	Factor analysis. Moran's I. SLM (spatial regression method, which can incorporate spatial dependency into the classical regression model). Monte Carlo estimation. Statistical significance at $p < 0.05$	GeoDa 0.9.5-I
Colak et al. (2015); PMID: 25619041	Geostatistical analysis of the relationship between heavy metals in drinking water and cancer incidence in residential areas in the Black Sea region of Turkey.	Black Sea region (Turkey)	Overall cancer incidence	17 heavy metal elements	No	Village/district	Water sources inside the villages/districts	Village/district	Kriging method. Linear regression analysis. Statistical significance at $p < 0.05$, $p < 0.01$	ArcGIS 10. SPSS 10
Hanchette et al. (2018); PMID: 30065203	Ovarian Cancer incidence in the U.S. and toxic emissions from pulp and paper plants: A geospatial analysis.	45 federal states and Washington D.C. (USA)	White females ovarian cancer incidence	Toxic air and water releases from pulp and paper mills	Only white females	County	ZIP code, county, and EPA region	County	Exploratory spatial data analysis: Moran's I and local indicator of spatial autocorrelation (LISA). Ordinary least squares (OLS) regression first for both the state- and county-level data. Spatial lag models for the state-level data. For the county-level data, GWR models.	ArcGIS 10.5; GeoDa

(continued)

Table 2 (continued)

Reference PMID	Title	Country/region	Dependent variable	Environmental factors	Socio-economic variables	Smallest area unit (dep. variable)	Smallest area unit (envir. factor)	Final smallest area unit considered	Methods	Software
Hendryx et al. (2012); PMID: 22471926	Permitted water pollution discharges and population cancer and non-cancer mortality: toxicity weights and upstream discharge effects in US rural–urban areas	Urban–rural areas (USA)	Mortality rates for cancer, kidney disease, and total non-cancer causes	Permitted toxic chemical pollutants in surface waters	College education rates, poverty rates, race/ethnicity percentages, rural–urban	County	County	County	Statistical significance at $p < 0.05$, $p < 0.01$, $p < 0.001$	Not found
Huang et al. (2013); PMID: 23575356	Cell-type specificity of lung cancer associated with low-dose soil heavy metal contamination in Taiwan: an ecological study.	Taiwan	Lung cancer incidence	7 heavy metals (As, Cd, Cr, Cu, Hg, Pb, Ni, Zn) concentrations in soil	Sex, age.	Townships	Townships	Townships	Descriptive statistics and examination for multicollinearity, followed by non-spatial and spatial analyses (GWR). Statistical significance at $p < 0.01$ Poisson regression models. Statistical significance at $p < 0.05$	SAS 9.13
Jian et al. (2017); PMID: 27713110	Associations between Environmental Quality and Mortality in the Contiguous United States, 2000–2005.	County by rural–urban continuum (USA)	All-cause mortality rate, heart disease, cancer, stroke	Environmental quality index (EQI)	Rural–urban continuum codes, percent of white population and the population density	County	County	County	Linear regression model to assess the average effects for the contiguous United States. Random intercept, random slope hierarchical model clustered	R 3.2.0 with the package lme4

Reference	Location	Outcome	Exposure	Covariates	Spatial unit (outcome)	Spatial unit (exposure)	Spatial unit (analysis)	Methods	Software
Lin et al. (2014); PMID: 24566045	Taiwan	Oral cancers age-standardized mortality rates	7 heavy metals (As, Cd, Cr, Cu, Pb, Ni, Zn) concentrations in soil	No	District	1 km × 1 km grid scale	1 km × 1 km grid scale	ATP Poisson kriging estimation. Anselin local Moran's I. Statistical significance at $p < 0.05$, $p < 0.001$	R
López-Abente et al. (2018a); PMID: 28155030	Spanish towns (Spain)	Mortality for 13 types of malignant tumors	Topsoil metal concentrations	Socio-demographic indicators: Population size, percentages of illiteracy, farmers, unemployment, average number of persons per household, mean income.	Town area (municipality)	Sampling locations	Cells 5 × 5 km	Kriging estimation. Factor analysis. BYM model with integrated nested Laplace approximations. Statistical significance at $p < 0.05$	R with the geoR, StatDA, and INLA packages
López-Abente et al. (2018b); PMID: 28847132	Galicia (Spain)	14 cancer sites incidence	Radon/Arsenic (in topsoil)	Socio-demographic indicators: population size, percentages of illiteracy, farmers, unemployment, average number of persons per household, mean income.	Town area (municipality)	Sampling locations	Cells 10 × 10 km	BYM model with integrated nested Laplace approximations. Statistical significance at $p < 0.05$	R with the INLA package

(continued)

Table 2 (continued)

Reference PMID	Title	Country/region	Dependent variable	Environmental factors	Socio-economic variables	Smallest area unit (dep. variable)	Smallest area unit (envir. factor)	Final smallest area unit considered	Methods	Software
Messier and Serre (2017); PMID: 27639278	Lung and stomach cancer associations with groundwater radon in North Carolina, USA.	North Carolina central Cancer registry (USA)	Stomach cancer and lung cancer incidence	Groundwater radon concentration (Bq/L)	Age, gender, race, residential tenure	Census tract	Census tract	Census tract	Negative binomial GLM with standard NB2 parameterization. Anselin local Moran's I. spatial autocorrelation of model residuals is assessed by examining a spatial covariance plot of the model standardized Pearson residuals. If present, a generalized estimating equation (GEE), which accounts for correlations between clusters and assumes no correlation within clusters, is implemented. Statistical significance at $p < 0.05$	R with the COUNT and GEE packages. BMElib numerical toolbox in MATLAB. Cluster and outlier analysis tool in ArcGIS 10.0.
Núñez et al. (2016); PMID: 27239676	Arsenic and chromium topsoil levels and cancer mortality in Spain.	Spanish towns (Spain)	Mortality for 27 types of malignant tumors	Arsenic and chromium (in topsoil)	Socio-demographic indicators: Population size, percentages of illiteracy, farmers, unemployment, average number of persons per household, mean income.	Town area (municipality)	Sampling locations	Town area (municipality)	Kriging estimation. Factor analysis. BYM model with integrated nested Laplace approximations. Statistical significance at $p < 0.05$	R with the INLA package

Núñez et al. (2017); PMID: 28108922	Association between heavy metal and metalloid levels in topsoil and cancer mortality in Spain.	Spanish towns (Spain)	Mortality for 27 types of malignant tumors	Topsoil metal concentrations	Socio-demographic indicators: Population size, percentages of illiteracy, farmers, unemployment, average number of persons per household, mean income.	Town area (municipality)	Sampling locations	Town area (municipality)	Kriging estimation. Factor analysis. BYM model with integrated nested Laplace approximations. Statistical significance at $p < 0.05$	R with the geoR, StatDA, and INLA packages
Ren et al. (2014); PMID: 25546281	Association between changing mortality of digestive tract cancers and water pollution: a case study in the Huai River Basin, China.	Huai River Basin (China)	Digestive cancer mortality	A series of frequency of serious pollution (FSP) indices including water quality grade (FSPWQG), biochemical oxygen demand (FSPBOD), chemical oxygen demand (FSPCOD), and ammonia nitrogen (FSPAN)	Gross domestic product	County	County	County	Linear correlation. Statistical significance at $p < 0.10$; $p < 0.05$, $p < 0.01$	Not declared
Roh et al. (2017); PMID: 28841521	Low-level arsenic exposure from drinking water is associated with prostate cancer in Iowa.	87 out of the 99 Iowa counties (USA)	White males prostate cancer incidence	Arsenic (in drinking water)	Poverty rate (only white males)	County	County	County	Spatial Poisson regression model. Anselin local Moran's I. Statistical significance at $p < 0.05$	SAS 9.4

(continued)

Table 2 (continued)

Reference PMID	Title	Country/region	Dependent variable	Environmental factors	Socio-economic variables	Smallest area unit (dep. variable)	Smallest area unit (envir. factor)	Final smallest area unit considered	Methods	Software
Saint-Jacques et al. (2018); PMID: 29089168	Estimating the risk of bladder and kidney cancer from exposure to low-levels of arsenic in drinking water, Nova Scotia, Canada.	Nova Scotia (Canada)	Bladder cancer and kidney cancer incidence	Arsenic (in drinking water)	Area-based composite indices of material and social deprivation	Set of continuous 25 km^2 cells	Set of continuous 25 km^2 cells	Set of continuous 25 km^2 cells	BYM model with integrated nested Laplace approximations. Statistical significance at $p < 0.05$	R with the disease mapping and INLA packages
Su et al. (2010); PMID: 20152030	Incidence of oral cancer in relation to nickel and arsenic concentrations in farm soils of patients' residential areas in Taiwan.	Taiwan	Oral cancers age-standardized mortality rates	8 heavy metals (As, Cd, Cr, Cu, Hg, Ni, Pb, Zn) in soil	Personal income, factory density, factory distribution and types of industry, and other socio-economic variables	Township/precinct	Township/precinct	Township/precinct	Step-wise multiple regression. Global Moran's I. Spatial models including conditional autoregressive model (CAR) and spatial simultaneous autoregressive (SAR) model. Statistical significance at $p < 0.05$	S-plus with spatial module
Van Leeuwen et al. (1999); PMID: 10597979	Associations between stomach cancer incidence and drinking water contamination with atrazine and nitrate in Ontario (Canada) agroecosystems, 1987–1991.	Ontario Cancer registry (Canada)	Age-standardized cancer incidence ratios: Stomach, colon, ovary, bladder, central nervous system, non-Hodgkin's lymphoma	Atrazine and nitrate in agroecosystems	Education level, income, occupation	Census sub-division (CSD)	Ecodistricts	Census sub-division (CSD)	Descriptive statistics and omnibus test. Least squares regression analysis. Global Moran's I. Statistical significance at $p < 0.25$, $p < 0.15$, $p < 0.05$	SPACESTAT

PMID PubMed identifier

Table 3 Articles classified as Type 2 and main characteristics

Reference; PMID	Title	Country/ region	Dependent variable	Environmental factors	Socio-economic variables	Smallest area unit (dep. variable)	Smallest area unit (envir. factor)	Final smallest area unit considered	Methods	Software
Dahl et al. (2013); PMID: 22569744	Is the quality of drinking water a risk factor for self-reported forearm fractures? Cohort of Norway.	Norway	Forearm fractures	Main water quality indicators.	Marital status; education level; urban–rural residence	Geographic coordinates	Geographic coordinates	Geographic coordinates	GAM. Statistical significance at $p < 0.05$	ArcGIS 9.3. STATA 11
Edwards et al. (2014); PMID: 24506178	Regional specific groundwater arsenic levels and neuropsychological functioning: a cross-sectional study.	Texas Alzheimer's research and care consortium (USA)	TARCC neuropsychology scores	Arsenic in groundwater	Age, gender, education	Region	Cells of 0.8 square miles	Region	Linear regression models. Statistical significance at $p < 0.05$	ArcGIS
McDermott et al. (2014); PMID: 24771409	Does the metal content in soil around a pregnant woman's home increase the risk of low birth weight for her infant?	South Carolina (USA)	Low birth weight	8 heavy metals (As, Ba, Cr, Cu, Pb, Mn, Ni, Hg) in soil	Maternal age and race; number of priorbirths	GIS coordinates	GIS coordinates	GIS coordinates	Multivariable GAM. Statistical significance at $p < 0.001$	ArcGIS9.3. R with the mgcv package

(continued)

Table 3 (continued)

Reference; PMID	Title	Country/ region	Dependent variable	Environmental factors	Socio-economic variables	Smallest area unit (dep. variable)	Smallest area unit (envir. factor)	Final smallest area unit considered	Methods	Software
Monrad et al. (2017); PMID: 28157645	Low-level arsenic in drinking water and risk of incident myocardial infarction: A cohort study.	Denmark	Myocardial infarction incidence	Arsenic in drinking water	Education level	Individual	Water supply area	Individual	Time-weighted average concentration. Evaluation of the exposure–response association by a cubic spline function with continuous first and second derivatives with 3 and 6 knots. Poisson GLM model. Statistical significance at $p < 0.05$	SAS (Lexis macro and PROC GENMOD procedure)

PMID PubMed identifier

Table 4 Articles classified as Type 3 and main characteristics

Reference; PMID	Title	Country/region	Dependent variable	Environmental factors	Socio-economic variables	Smallest area unit (dep. variable)	Smallest area unit (envir. factor)	Final smallest area unit considered	Methods	Software
Banning and Benfer (2017); PMID: 28820453	Drinking water uranium and potential health effects in the German Federal State of Bavaria.	Bavaria federal state (Germany)	Cancer and other diseases incidence	Uranium in drinking water	No	County	Municipality	Counties	Municipality concentration level used for the entire county, where available. Then classification in groups. Pearson correlation between SIR and concentration groups. Statistical significance at $p < 0.05$, $p < 0.01$	ArcGIS 10.1. SPSS
Cech et al. (1987); PMID: 3610447	Health significance of chlorination byproducts in drinking water: The Houston experience.	Houston, Texas (USA)	Mortality by urinary tract cancer, respiratory cancers, non-cancer respiratory causes	Trihalomethanes in drinking water	Gender, age, race	Census tract	Census tract	Census tract	Trends compared to pollutant concentration. Statistical significance at $p < 0.05$, $p < 0.01$	Not declared

(continued)

Table 4 (continued)

Reference; PMID	Title	Country/ region	Dependent variable	Environmental factors	Socio-economic variables	Smallest area unit (dep. variable)	Smallest area unit (envir. factor)	Final smallest area unit considered	Methods	Software
Collman et al. (1988); PMID: 3198278	Radon-222 concentration in groundwater and cancer mortality in North Carolina.	North Carolina (USA)	Deaths from cancers of the nasal cavities, oro-, naso-, and hypopharynx, larynx, esophagus, stomach, colon, breast, bone, and the four major types of leukemia	Radon in public water supply	No	County	County	County	Relative risk by radon concentration. Statistical significance at $p < 0.05$	Not declared
Crump et al. (1987); PMID: 3591777	Cancer incidence patterns in the Denver metropolitan area in relation to the rocky flats plant.	Rocky flats, Colorado (USA)	Various cancers incidence	Plutonium in soil	Gender, age	Census tract	Census tract	Census tract	Bivariate analyses. Mantel-Haenszel test. Statistical significance at $p < 0.05$, $p < 0.01$, $p < 0.001$	Not declared
Dreiher et al. (2005); PMID: 16330453	Non-Hodgkin's lymphoma and residential proximity to toxic industrial waste in southern Israel.	Southern Israel	Non-Hodgkin's lymphoma incidence and survival	Toxic industrial waste	Gender, age, ethnicity, occupation	14-kms. Radius near the pollution source	14-kms. Radius near the pollution source	14-kms. Radius near the pollution source	GIS standardized rates. Kaplan–Meier method. Cox proportional hazard regression. Statistical significance at $p < 0.05$	MapInfo and not declared

Grilc et al. (2015); PMID: 27646727	Drinking water quality and the geospatial distribution of notified gastrointestinal infections.	Slovenia	Acute gastrointestinal infections incidence	Fecal contamination of water supply system	No	Water supply zone	Water supply zone	Water supply zone	Classification of contaminated zones in three groups. Comparison with incidence in the same areas, computing the RRs. Statistical significance at $p < 0.05$	ArcGIS 10. Oracle 11 g
Richmond et al. (1987); PMID: 3616722	Colorectal cancer mortality and incidence in Campbell County, Kentucky.	Campbell County, Kentucky (USA)	Colon-rectum cancer incidence and mortality	Trihalomethanes in kitchen tap water	Gender, age, occupation	Census block	Census block	Census block	SIR and SMR compared to pollutant concentration. Statistical significance based on the Poisson distribution method of Bailar and Ederer. Statistical significance at $p < 0.05$	Not declared
Sánchez-Díaz et al. (2018); PMID: 30423874	Geographic analysis of motor neuron disease mortality and heavy metals released to Rivers in Spain	Spanish rivers	Deaths from motor neuron disease	Arsenic, cadmium, copper, chromium, mercury, lead, zinc in waters	No	Municipality	20 kms. of the rivers section from the emission point	Municipality	Log-linear models (Poisson link function). Statistical significance at $p < 0.05$, $p < 0.001$	Stata. ArcGIS

(continued)

Table 4 (continued)

Reference; PMID	Title	Country/ region	Dependent variable	Environmental factors	Socio-economic variables	Smallest area unit (dep. variable)	Smallest area unit (envir. factor)	Final smallest area unit considered	Methods	Software
Thorpe and Shirmohammadi (2005); PMID: 16291529	Herbicides and nitrates in groundwater of Maryland and childhood cancers: a geographic information systems approach.	Maryland (USA)	4 childhood cancers incidence	Herbicides and nitrates	Gender, age, race	ZIP code	ZIP code	ZIP code	Cluster analysis. Contingency tables with chi-square analysis. Statistical significance at $p < 0.05$	ArcView 3.1. Spatial analyst 1.1. SaTScan 2.1. GraphPad prism 3.02

PMID PubMed identifier

Table 5 Articles classified as Type 4 and main characteristics

Reference; PMID	Title	Country/region	Dependent variable	Environmental factors	Socio-economic variables	Smallest area unit (dep. variable)	Smallest area unit (envir. factor)	Final smallest area unit considered	Methods	Software
Christian et al. (2011); PMID: 22043094	Exploring geographic variation in lung cancer incidence in Kentucky using a spatial scan statistic: Elevated risk in the Appalachian coal-mining region.	Kentucky (USA)	Lung cancer incidence	Coal mining waste and cigarette smoking	Gender, age	Circle areas	Circle areas	Circle areas	Discrete Poisson model. Monte Carlo simulation. Statistical significance at $p < 0.01$	SaTScan. ArcGIS 9.3
Cui et al. (2019); PMID: 30836673	Spatiotemporal variations in gastric Cancer mortality and their relations to influencing factors in S County, China	S County (China)	Gastric cancer mortality	Surface water quality	Population density, GDP	2x2 kms. Grid squares	2x2 kms. Grid squares	2x2 kms. Grid squares	Anselin local Moran's I. hot spot analysis. GeoDetector. Statistical significance at $p < 0.05$	ArcGIS 10.2. GeoDetector
Dai and Oyana (2008); PMID: 18939976	Spatial variations in the incidence of breast cancer and potential risks associated with soil dioxin	The Bay, Midland, and Saginaw counties, Central	Breast cancer incidence	Dioxin in soil	Age	ZIP code	ZIP code	ZIP code	Evaluation of soil dioxin contamination by using descriptive statistics and the SOM algorithm.	SOM toolbox. MatLab 7.1. ArcGIS 9.2. SatScan 7.0

(continued)

Table 5 (continued)

Reference; PMID	Title	Country/region	Dependent variable	Environmental factors	Socio-economic variables	Smallest area unit (dep. variable)	Smallest area unit (envir. factor)	Final smallest area unit considered	Methods	Software
	contamination in Midland, Saginaw, and Bay Counties, Michigan, USA.	Michigan (USA)							Evaluation of the association between breast cancer rates and the ZIP codes by estimating the odds ratio and their corresponding 95% confidence intervals. Cluster detection using Kulldorff's spatial and space-time scan statistics and genetic algorithms for spatial and space-time clustering. Statistical significance at $p < 0.05$, $p < 0.01$, $p < 0.001$	

Fei et al. (2018); PMID: 29679198	The association between heavy metal soil pollution and stomach cancer: a case study in Hangzhou City, China	Hangzou city (China)	Stomach cancer incidence	Heavy metals in soil	Gender	Township	Sampling points	Township	Spatial distribution of incidence tested by Global Moran's I. GeoDetector. Hotspot analysis for environmental factor's cluster. Kriging's interpolation. Statistical significance at $p < 0.01$	GeoDetector
Guajardo and Oyana (2009); PMID: 20049167	A critical assessment of geographic clusters of breast and lung cancer incidences among residents living near the Tittabawassee and Saginaw Rivers, Michigan, USA.	The Bay, Midland, and Saginaw counties, Central Michigan (USA)	Breast and lung cancer incidences	Various pollutant and pollutants	Median household income, race, percent of native born, education level, percent of population residing at the same address in 1995	ZIP code	ZIP code	ZIP code	Preliminary GIS analysis. Odds ratio statistics. Stepwise discriminant function analysis. Ordinary Kriging. Anselin Local Moran's I. Turnbull's method. Bithell's linear risk score test. Lawson and Waller score test. Statistical significance at $p < 0.05$, $p < 0.01$	ArcGIS 9.2. ArcView 3.3. GeoDa 0.95i. ClusterSeer 2.0 and TerraSeer's STIS 1.6. Excel. SPSS 17.0

(continued)

Table 5 (continued)

Reference; PMID	Title	Country/region	Dependent variable	Environmental factors	Socio-economic variables	Smallest area unit (dep. variable)	Smallest area unit (envir. factor)	Final smallest area unit considered	Methods	Software
Nieder et al. (2009); PMID: 19450849	Bladder cancer clusters in Florida: Identifying populations at risk.	Florida (USA)	Bladder cancer incidence	Arsenic in water	Race/ethnic categories, census derived poverty status at the block group level, census derived county-level urban/rural residence	Census block	Census block	Census block	Multivariate logistic regression. Statistical significance at $p < 0.05$, $p < 0.001$	ArcGIS 9.0. SaTScan 5.0. SPSS 11.0.1
Selvin et al. (1987); PMID: 3476785	Spatial distribution of disease: Three case studies.	Rocky flats, Colorado; Contra Costa County, California; Santa Clara County (California)	Lung cancer and leukemia incidence	Industrial facilities as proxy of pollution (pollutants not specified)	Gender, age, race	Census tract	Facility's position and distance from the cases	Census tract	Cluster analysis. Statistical significance at $p < 0.05$	Not declared

PMID PubMed identifier

3 Results

The PubMed search identified 694 articles. In *Phase 1* of the revision process, the reviewers agreed over 88% of the articles (considering both accepted and rejected articles). At this stage, 122 articles resulted eligible to be included in *Phase 2*. The complete read-through of 122 articles in *Phase 2* lead to an agreement of 61% among the four reviewers. After *Phase 3* of the revision, 40 articles were included in the review and classified as shown in Tables 2, 3, 4, and 5. Twenty of the articles referred to studies conducted in North America, 11 in Europe, 7 in Asia, and 2 in South America. The majority of articles (33 of 40) analyzed cancer incidence or mortality rates as outcome indicators. As for contaminants, 20 and 12 articles, respectively, analyzed pollutants in water and soil, while 7 articles analyzed pollutants in both elements and 1 article reported the results of applying the Environmental Quality Index to overall and by-cause mortality.

The statistical methods across the studies were quite diverse but could be grouped into specific families: descriptive analysis, data reduction procedures (factor analysis, cluster analysis), Moran's I and Kriging method for spatial interpolation, spatial regression analysis, various kinds of GLM regression models (often Poisson regression models), general additive models, Bayesian models with or without integrated nested Laplace approximations, and Monte Carlo estimations.

The authors of the studies we considered tested their results' statistical significance using different techniques; over half of the papers however did not report how they tested it (22). The remaining 18 articles used t-test (3 articles), chi-square, F-test, likelihood ratio test (2 articles for each test), Z score, Kruskal–Wallis test, Getis–Ord Gi statistic, Kulldorff's spatial and space-time scan statistics, Lawson and Waller score test, Mantel–Haenszel test and contingency table test, Poisson distribution method of Bailar and Ederer, Taylor series variance estimates, and a parametric bootstrap on testing for RR < 1.1 (1 article for each test). Statistical significance thresholds (p values) were reported in the Methods column of Tables 2, 3, 4, and 5.

As to packages, ArcGIS/ArcView and R are those principally used (14 articles each). See Fig. 1 for an overview of software use across the different studies.

The articles were synthetically classified in four groups, as shown in Table 1.

3.1 Type 1: Regression with Data by Geographical Area

Twenty articles (50%) were classified as belonging to this group (Table 2). Authors used different kinds of regression models to explore the relationship between health outcomes (dependent variable, e.g. cancer incidence or mortality), environmental factors, and any other covariate (e.g., socio-economic indicators) by geographical area. The geographic unit used to collect information on health outcome and pollution did not always match.

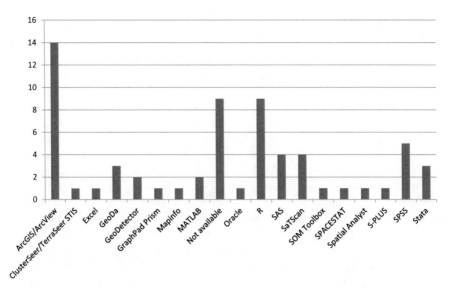

Fig. 1 Frequency of statistical software packages used

Eleven of the articles considered the same geographic areas for pollutants and for health outcomes (the areas either coincided or were very similar, e.g. districts, townships, etc.). In nine of the studies, information on pollution was collected using smaller geographic units than those used for health outcomes and addressed data misalignment in different ways. In five articles, authors applied the Kriging interpolation method (a Gaussian process regression) to extend the information collected at the pollution sources to the areas considered for epidemiologic population data (Colak et al. 2015; Lin et al. 2014; López-Abente et al. 2018a; Núñez et al. 2016, 2017). Authors of the remaining four articles used different approaches: the study by Hanchette et al. (2018) on the potential effects of toxic water releases on ovarian cancer incidence developed a combination of ordinary least squares (OLS) regression models and geographical weighted regression (GWR) models, corrected by spatial lag models (after testing the presence of local spatial autocorrelation by Local Indicators of Spatial Association – LISA – model). The study by López-Abente et al. (2018b) on the relationship between the presence of arsenic and radon in topsoil and 14 cancer sites incidence followed a Gaussian approach, which considered a Matérn Gaussian field approximated using the stochastic partial differential equation method for the environmental covariate. The study by Aballay et al. (2012) used a two-level model to estimate the effects of pollution in sampling points to the whole districts. Here, aquifer pollution was included as a random intercept and the misalignment was corrected by adaptive quadrature method. Finally, the study by Van Leeuwen et al. (1999) on atrazine and nitrate in drinking water and stomach cancer incidence determined mean contamination levels for each ecodistrict and data source; means were then proportionally combined to be associated to the population they represented.

Almost all 20 articles applied a preliminary exploratory spatial analysis using Moran's I for testing spatial autocorrelation after data geo-coding procedures.

As for regression models, the majority of studies incorporated different aspects of socio-economic, demographic, and life styles of the population under study in the evaluation of environmental exposure on the considered outcome. Only three articles (Chiang et al. 2010; Colak et al. 2015; Lin et al. 2014) did not correct the effects of such variables.

The choice of regression model varied between studies. Five studies applied the Besag, York, and Mollie's (BYM) model with integrated nested Laplace approximation. This is a convenient way to obtain approximations to the posterior marginal figures for parameters in Bayesian hierarchical models when the latent effects can be expressed as a Gaussian Markov random field, as it is defined in these works (López-Abente et al. 2018a, 2018b; Núñez et al. 2016, 2017; Saint-Jacques et al. 2018). Almost all other articles relied on different kinds of regression models, whether Poissonian or not, and considered the effects of spatial autocorrelation on the basis of the results of the Moran's I. Only the study by Armijo and Coulson (1975) on the relationship between stomach cancer mortality and presence of nitrate in drinking water and nitrogen fertilizers relied on bivariate correlation without considering the spatial autocorrelation term.

For the large part, two different kinds of software packages were used in these works, sometimes jointly, sometimes on their own. These were packages for geo-coding and study of spatial effects and model development.

R (S-Plus in one case) with its several packages was the most frequently used, because it made it easier to support geo-coding and analysis of spatial effects and to implement results in the Bayesian or regression models (Jian et al. 2017; Lin et al. 2014; López-Abente et al. 2018a, 2018b; Núñez et al. 2016, 2017; Messier and Serre 2017; Saint-Jacques et al. 2018; Su et al. 2010). Three studies relied on SAS for the versatility in adapting script adequate to combine the two aspects, as for R (Bulka et al. 2016; Huang et al. 2013; Roh et al. 2017). ArcGIS for spatial geo-coding and analysis was used in combination with other software packages (GeoDa and SPSS) in two studies (Colak et al. 2015; Hanchette et al. 2018); Stata, GeoDA, and Spacestat were used separately with few specific modules in 3 older studies (Aballay et al. 2012; Chiang et al. 2010; Van Leeuwen et al. 1999), while it was not possible to identify the software used for three of the articles (Armijo and Coulson 1975; Hendryx et al. 2012; Ren et al. 2014).

3.2 Type 2: Regression Models at Individual Level

Four articles (10%) pertained to this group (Table 3). The relationship between health condition/outcome and environmental factor was analyzed by regression models at individual level. The methodological interest was focused on the definition of an individual value for the environmental factor.

The geographic level for this group was mainly the individual, geo-coded at the geographic coordinates of residence (Dahl et al. 2013; McDermott et al. 2014; Monrad et al. 2017). Only Edwards et al. (2014) used the region of residence for attribution of exposure, but it mainly relied on the Texas Alzheimer's Research and Care Consortium (TARCC) neuropsychology scores at individual level for the analysis. Pollution sources were geo-coded at the same level (Dahl et al. 2013; McDermott et al. 2014) or at a slightly larger scale (water supply area (Monrad et al. 2017) or cells of 0.8 mile^2 (Edwards et al. 2014)). Pollution was then reported to the individual level by time-weighted average concentration and binary classification of exposure (McDermott et al. 2014, Monrad et al. 2017), stratification of exposure in groups (Dahl et al. 2013), or by attribution of the pollutant concentration in any cell to the corresponding person (Edwards et al. 2014).

All four studies used demographic and socio-economic covariates for correcting the environmental effects in different regression models such as linear regression (Edwards et al. 2014), Poisson generalized linear model (Monrad et al. 2017), and generalized additive model (Dahl et al. 2013, McDermott et al. 2014).

As with software packages, three articles used ArcGIS for data geo-coding (Dahl et al. 2013; Edwards et al. 2014; McDermott et al. 2014) and two of them combined other software packages for the models (R and Stata) (Dahl et al. 2013, McDermott et al. 2014). The study by Monrad et al. (2017) used SAS with specific procedures.

3.3 Type 3: Exposure Intensity Threshold Values for Evaluating Health Outcome

Nine articles (22.5%) were grouped as Type 3 (Table 4), characterized by a hiatus between the study of environmental factors and the distribution of health outcomes. The environmental factor, often detected as punctual source, was recoded as a categorical variable and the considered geographic areas were classified on the basis of the values/characteristics of such environmental categorical variable. The health outcome was analyzed at the same or larger area level. Therefore, threshold values for exposure intensity were computed in order to define cut-off points for evaluating trends in the health outcome variable to study the influence of the environmental factor.

In almost every article, the geographic areas considered for health outcomes, environmental factor, and other covariates were homogeneous; a few differences existed only in the studies by Banning and Benfer (2017) (county vs. municipality) and by Sánchez-Díaz et al. (2018) (municipality vs. river sections of 20 kms). In the first case, the pollutant concentration level in the municipality was extended to the entire county; in the second case, the case distance from the pollution source was considered as an independent variable of exposure in the final model.

Methods were not homogeneous due to differences in cut-off definitions and in the evaluation of their statistical significance with respect to the considered health outcome. Five articles also considered demographic and socio-economic characteristics of the population (Cech et al. 1987; Crump et al. 1987; Dreiher et al. 2005; Richmond et al. 1987; Thorpe and Shirmohammadi 2005).

ArcGIS and MapInfo were used for geo-coding, defining the cut-off points for the environmental factor and some spatial analysis (Banning and Benfer 2017; Dreiher et al. 2005; Grilc et al. 2015; Sánchez-Díaz et al. 2018; Thorpe and Shirmohammadi 2005); SPSS, SaTScan, and Stata for analyzing the potential correlation. In four articles, the used software packages were not declared (Cech et al. 1987; Collman et al. 1988; Crump et al. 1987; Richmond et al. 1987).

3.4 Type 4: Distance between Pollutant Source and Health Outcome Clusters

Seven articles (17.5%) belonged to this group (Table 5). No association between pollutants and health outcomes was considered in the first phase of these studies. Initially, they identified separately clusters of areas or people generated by the analysis of the considered health outcomes and environmental pollution geographic clusters obtained by considering environmental factors and their potential emission sources. As a second step, authors performed a comparison between health outcome geographic clusters and environmental pollution geographic clusters to evaluate their superimposition or proximity.

In five articles, the geographic areas considered for pollutants and health outcomes coincided (Christian et al. 2011; Cui et al. 2019; Dai and Oyana 2008; Guajardo and Oyana 2009; Nieder et al. 2009), thus reducing issues linked with the estimation of pollution concentration in areas wider than the one observed. The study by Selvin et al. (1987) on the relationship between leukemia, lung cancer incidence and industrial waste pollution used the distance between potential emission source and centroid of cases' residence census tract. This indicator became the factor connecting the cluster of disease with the cluster of pollution. The study by Fei et al. (2018) used the township of residence to geo-position the cases and a number of pollution sampling points in Hanghzou city; the authors joined this information with the hotspot analysis and the Kriging interpolation method so as to extend the pollutants concentration to the townships.

All articles used Moran's I as main indicator for evaluating spatial autoregression effects both on the environmental factors and the health outcomes. Also demographic, socio-economic, and life styles factors were considered in every work.

Different methods and techniques were used for the purpose of identifying environmental and health outcome clusters. These included classical cluster analysis (Selvin et al. 1987), Monte Carlo simulation and hypothesis testing for the identification of excess risk clusters (Christian et al. 2011; Nieder et al. 2009). Moreover, statistical analyses were performed by different score tests after combination of GIS and spatial techniques (Guajardo and Oyana 2009; Dai and Oyana 2008) and finally the quite recent GeoDetector, a spatial stratification statistical technique (Cui et al. 2019; Fei et al. 2018).

The studies in this group used a variety of software packages to address every specific issue, this is due to the peculiarity of these studies (all of them quite exploratory of not yet well-defined local situations). ArcGIS (in its various version) was used in almost every article for geo-coding and for some spatial analysis; SaTScan allowed to work in terms of "circles" of different, varying radius (Christian et al. 2011; Dai and Oyana 2008; Nieder et al. 2009); GeoDA, SPSS, ClusterSeer, TerraSeer and a few adaptable packages such as MATLAB, SOM Toolbox, and GeoDetector were used for finding and evaluating the statistical significance of the clusters (Cui et al. 2019; Dai and Oyana 2008; Fei et al. 2018; Guajardo and Oyana 2009).

4 Discussion

Our WASABY project herewith identifies and points out a number of public health studies that, regardless of their aims, may be of interest for the investigation of the relationship between environmental factors and health outcomes using available data.

The issue of exposure assessment has been investigated for over 40 years (the oldest study selected in this review dates 1975) and during this period significant changes were introduced in terms of the pollutants considered or in terms of the health outcomes analyzed. Innovations covered new technologies to measure pollutants, statistical methodologies to assess exposure, and software and hardware progress. These changes allowed to develop and apply geo-coding and statistical methods for the reduction of the ecological bias when considering the relationships among individuals, geographic areas, pollutants, and health outcomes (Woods et al. 2005).

More complex models for interpolation and analysis have become available with the development of software and hardware allowing for increased computation power. Most of the studies we considered were developed after the first decade of the twenty-first century (29 studies were published after 2009) when tools for spatial analysis and representation were greatly developed and made more user-friendly, thanks to the introduction of more powerful processors. This was particularly true for spatial interpolation and estimation of multifactor effects which used to be applied to large datasets. As an example, the Intel Core microprocessors (I3-I7) became available in 2010 offering superior computational power.

Following the growing demand for these types of studies, new packages and user-interfaces for free programs (e.g., R) and scripts for commercial programs (e.g., SAS, Stata) were developed. Specifically, procedures such as the Kriging interpolation, the computation of Moran's I, the application of Poisson linear regression, or INLA models became more accessible after the introduction of new tools and improvements. Geographic representation programs markedly improved including internal tools for simple and more specific statistic analysis as well as more user-friendly interfaces thus widening the audience of users.

Criteria for software choice naturally include availability of specific tools/scripts for a) management and linkage of large datasets, b) spatial interpolation and advanced analysis (like the INLA models in R), c) geographic representation, and d) for cost. In consideration of the above-mentioned criteria, R is often considered the best choice thanks to the extension of available tools that allow to develop all procedures for free.

Commercial programs such as Stata and SAS offer more user-friendly interfaces at higher costs. For this reason, they are chosen by virtue of the availability and quality of the scripts.

As to geographic representations, ArcGIS (commercial), QGIS, and SaTScan (free) appear to be the best choice, owing to their usability, connection with online map sources, and presence of internal tools for both simple and more sophisticated spatial analyses.

A major merit of our study is the identification and critical evaluation of published articles on the topic by four individual researchers under standardized criteria and methods. In our review we highlighted some of the most recent studies, methodologies, and techniques able to define the smallest available units of observation (e.g., the census tract or specific small territorial cells defined in each research). This improved the estimation reliability of the effects on health due to the exposure to pollutants and other factors when transferring considerations from "area" to "person" (Lillini and Vercelli 2019), in compliance with EU privacy legislation on analyses at the individual level.

Our work does not intend to offer a comprehensive overview of methodologies for ecological environmental studies on water and soil pollution in relation to public health, as relevant articles on these issues might have been missed out as a result of the term search criteria. However, we hope to have intercepted most of the main relevant methodologies and techniques.

Another limitation of our study is the exclusion of non-English language papers. A number of articles written in Chinese, Italian, Russian, and Spanish were not considered in this work due to sub-optimal readability (Chinese and Russian ones) and comprehensibility (Spanish ones) as well as to enhance the possibility of reaching a wide audience.

The methods reported in this review are appropriate for research on water and soil pollution data, as detailed in the rationale of the WASABY project; for this reason, they could not be generalized to other environmental risk factors, such as air pollutants.

Finally, most of the considered works shared the cross-sectional study design, as expected.

Overall, our analysis shows a wide variation of valid and reliable methods and techniques. It is not possible to identify a "gold standard" because of the peculiarity of every situation. On the other hand, when approaching such issues, scholars may identify the research experiences that best fit the situation they are approaching to investigate, apply all corresponding procedures, and adapt them to the specific situation they are facing.

Here, we wish to give our insight on the use of different statistical models so as to provide some advice for choosing the best option for different research aims. First, researchers will have to choose whether to opt for a frequentist or Bayesian approach. This choice is both theoretical and practical (Samaniego 2010).

The frequentist approach assumes one's measurements are enough to state something meaningful. Probability is defined in terms of limiting frequency of occurrence of an event, it assumes that there are true values of the model parameters and it computes the parameters point estimates. In the Bayesian approach, data are supplemented with additional information in the form of a prior probability distribution. The prior belief about the parameters is combined with the data's likelihood function according to Bayes theorem, in order to yield the posterior belief about the parameters. Probability is the degree of belief on the occurrence of an event, only data are real and there are no true values of parameters as such, apart from the fact that a number of values are more probable than others.

Most frequently used models are linear ones, e.g., Poisson regression or general additive models (GAMs), Besag York Mollié (BYM) models with or without integrated nested Laplace approximation (INLA).

Poisson regression seems appropriate when the dependent variable is a count, the events must be independent, but the probability per unit time of events is understood to be related to covariates. Poisson regression is also appropriate for rate data, where the rate is a count of events divided by part of a given unit's exposure (a particular unit of observation). Event rates may be calculated as events per unit time, which allows the observation window to vary for each unit. Here, exposure is respectively unit area, person−years, and unit time (Tutz 2011).

GAMs are a class of statistical models in which the usual linear relationship between the response and predictors is replaced by several non-linear smooth functions to model and capture the non-linearity of data. These are flexible techniques that help to fit linear models which can either be linearly or non-linearly dependant on several predictors. The latter characteristic makes them very useful and reliable to identify and describe non-linear relationships between response and predictors. There are at least three good reasons for using GAM: interpretability, flexibility/automation, and regularization. When the model contains non-linear effects, GAM provides a regularized and interpretable solution, while other methods generally lack at least one of these three features (Hastie and Tibshirani 1990).

BYM model is a Bayesian hierarchical model based on a conditional autoregressive (CAR) model for spatial random effects. In the CAR model, spatial dependence is expressed conditionally: given the values in all other areas, it requires that the random effect in an area depends only on a small set of neighboring values. An essential aspect of the BYM model and its extensions is the specification of the neighborhood structure for the areas. This is quite flexible and it may be arbitrarily defined. It is based on adjacency relationships of the geographical areas (or disjoint geographical areas with the needed correction) (Rodrigues and Assunção 2012). BYM is useful to investigate the underlying relative risks of a disease observed on joint or disjoint geographical areas. On the other end, however, it needs a stable and quite homogeneous definition of the geographical units and outcomes, and covariates must be defined at the same geographical level or they should be interpolated at such level.

In some studies, BYM models were developed along with INLA, which relies on a combination of analytical approximations and efficient numerical integration schemes to achieve highly accurate deterministic approximations to posterior quantities of interest. The main benefit for using INLA instead of Markov chain Monte Carlo (MCMC) techniques is computation. INLA is fast even for large and complex models. Moreover, INLA is a deterministic algorithm and does not suffer from slow convergence and poor mixing (Rue et al. 2009).

A common aspects considered in spatial analysis is the spatial autocorrelation, i.e. the co-variation of properties within geographic space. Characteristics at proximal locations appear to be correlated, either positively or negatively. The spatial autocorrelation problem violates the condition of standard statistical techniques that assume independence among observations (Knegt et al. 2010). Spatial analysis models correct spatial autocorrelation with different techniques. In the studies analyzed by this review, spatial autocorrelation is measured by Moran's I, a correlation coefficient that measures the overall spatial autocorrelation of the data set. In other words, it measures how one object is similar to others surrounding it. If objects are attracted (or repelled) by each other, it means that the observations are not independent. The standardized version of Moran's I enables to compare the significant spatial patterns of different or same variables with different calculating parameters and it should be chosen as the preferred test (Getis 2010).

Another observation regards the convergence of the geographical level at which the data is collected. In several cases, health outcomes, environmental variables, and other covariates (e.g., socio-economic data) are collected at the same geographic level (e.g., municipality, census tract, etc.). In other cases areas do not coincide due to the different availability and data characteristics in the selected sources. When this is the case, interpolating methods must be applied to reduce territorial bias.

Many of the articles considered in our study used Kriging regression as the preferred method of interpolation so values are modeled by a Gaussian process governed by prior covariances. Under suitable assumptions on the priors, Kriging gives the best linear unbiased prediction of the intermediate values. Interpolating methods based on other criteria such as smoothness (e.g., smoothing spline) may not yield the most likely intermediate values. There are two Kriging methods: ordinary and universal. Ordinary Kriging is the most general and widely used of the Kriging methods. Here, the constant mean is assumed as unknown. Universal Kriging assumes that there is an overriding trend in the data which can be modeled by a deterministic function, a polynomial. Universal Kriging should only be used when you know there is a trend in your data and you can give a scientific justification to describe it. The method can be used where spatially-related data has been collected and estimates of "fill-in" data are desired in the locations (spatial gaps) between the actual measurements (Oliver and Webster 1990).

Another example of the research choice to be made in these types of studies is the use of socio-economic information as a correction of the effects of the environmental conditions on health outcomes. This correction should always be considered in such type of studies if the socio-economic data are available and reliable. This is because there is a relevant relationship between these characteristics, the probability of living in areas where exposure to pollution is significant, and the health condition of the considered population (see, for instance, the founding work of Dolk et al. 1995, or the more recent Pannullo et al. 2016). When socio-economic characteristics are not considered, the bias is a more superficial description of the interested population and it is possible to lose relevant indication to better address health policy actions. It is, therefore, advisable to collect socio-economic data, at least at small geographic area (e.g., Census Tract, Woods et al. 2005). In many countries, ecological socio-economic deprivation indices at such geographic level are already available (e.g., Guillaume et al. 2016; Caranci et al. 2010).

5 Conclusion

This review represents a useful tool for cancer registries, health institutions, and environmental agencies that are interested in territorial monitoring, health or environmental surveillance. We provide suggestions on methods, techniques, and tools which may be applied in studies that investigate disease clusters and environmental exposure. In this perspective, the study contributes to optimize the use of public health resources.

Acknowledgments This document was produced by "WASABY – Water And Soil contamination and Awareness on Breast cancer risk in Young women", which received funding from the European Commission as part of the framework of the Health Program (Contract No. PP-2-2016 with the DG SANTE Directorate C). *Wasaby Working Group*: FRANCE. *Registre général des cancers de Poitou-Charentes:* Gautier Defossez. *Registre des tumeurs de Loire-Atlantique et Vendée:* Florence Molinie; Anne Cowpli-Bony. *Registre général des cancers de Haute Vienne:* Tania Dalmeida. *Registre général des tumeurs du Calvados:* Anne Valérie Guizard. *Registre des cancers de la Manche:* Simona Bara. *Registre du cancer de la Somme:* Bénédicte Lapotre-Ledoux. *Registre général des cancers de Lille et de sa région:* Sandrine Plouvier. *Registre des cancers du Bas-Rhin:* Michel Velten. *Registre des cancers du Haut-Rhin:* Séverine Boyer. *Registre des tumeurs du Doubs:* Anne Sophie Woronoff. *Registre des cancers du sein et des cancers gynécologiques de Côte-d'Or:* Patrick Arveux. *Registre du cancer de l'Isère:* Marc Colonna. *Registre des tumeurs de l'Hérault:* Brigitte Tretarre. *Registre des cancers du Tarn:* Pascale Grosclaude. *Registre général des cancers de Guadeloupe:* Jacqueline Deloumeaux. *Registre général des cancers de Martinique:* Clarisse Joachim. *Registre des cancers de Gironde:* Gaelle Coureau. *Université de Caen:* Josephine Bryere ITALY. *Analytical Epidemiology and Health Impact Unit, Fondazione IRCCS "Istituto Nazionale dei Tumori", Milano:* Simone Bonfarnuzzo, Milena Sant. *Registro Tumori dell'Alto Adige:* Guido Mazzoleni; Fabio Vittadello. *Registro Tumori ASL Napoli 3 Sud:* Mario Fusco; Maria Francesca Vitale; Valerio Ciullo. *Registro Tumori di Palermo e provincia:* Rosanna Cusimano; Walter Mazzucco; Maurizio Zarcone. *Registro Tumori di Parma:* Maria Michiara; Paolo Sgargi. *Registro Tumori ASP Ragusa-Caltanisetta:* Carmela Nicita; Rosario Tumino. *Registro Tumori della Provincia di Siracusa:* Ylenia Dinaro; Francesco Tisano. *Registro Tumori di Trapani e*

Agrigento: Giuseppa Candela; Tiziana Scuderi. *Registro Tumori di Trento:* Roberto Rizzello; Silvano Piffer. *Registro Tumori Umbria:* Fabrizio Stracci; Fortunato Bianconi. *Registro Tumori di Varese:* Giovanna Tagliabue. LITHUANIA. *Lithuanian Cancer Registry:* Ieva Vincerzevskiene. POLAND. *Polish National Cancer Registry:* Joanna Didkowska; Urszula Wojciechowska; Krzysztof Czaderny. *Greater Poland Cancer Registry:* Łukasz Taraszkiewicz; Maciej Trojanowski. *Kielce Cancer Registry:* Pawel Macek. *Masovia Cancer Registry:* Urszula Sulkowska. *Silesia Cancer Registry:* Marcin Motnyk. *Subcarpatian Cancer Registry:* Monika Gradalska-Lampart. PORTUGAL. *Registo Oncológico Regional do Norte:* Luis Antunes; Jéssica Rodrigues. *Registo Oncológico Regional - Zona Centro:* Joana Antunes Lima Bastos; Margarida Ornelas. SPAIN. *Registro de Cáncer de Euskadi-CIBERESP:* Arantza Lopez de Munain; Nerea Larrañaga. *Registro de Tumores de Castellón-Valencia:* Paloma Botella; Consol Sabater Gregori. *Registro de Cáncer de Girona:* Marc Saez; Rafael Marcos-Gragera. *Registro de Cáncer de Granada, EASP, CIBERESP, ibs.GRANADA, UGR:* Maria Jose Sanchez-Perez; Miguel Rodriguez-Barranco. *Registro de Cáncer de Murcia-CIBERESP:* Monica Ballesta-Ruiz; Maria Dolores Chirlaque. *Registro de Cáncer de Navarra-CIBERESP:* Eva Ardanaz; Marcela Guevara. NORTHERN IRELAND (UK). *Northern Ireland Cancer Registry:* Anna Gavin; David Donnelly.

Declaration of Competing Financial Interests (CFI) The authors declare they have no actual or potential competing financial interests.

References

Aballay LR, Díaz Mdel P, Francisca FM, Muñoz SE (2012) Cancer incidence and pattern of arsenic concentration in drinking water wells in Córdoba, Argentina. Int J Environ Health Res 22 (3):220–231. https://doi.org/10.1080/09603123.2011.628792

Armijo R, Coulson AH (1975) Epidemiology of stomach cancer in Chile--the role of nitrogen fertilizers. Int J Epidemiol 4(4):301–309. https://doi.org/10.1093/ije/4.4.301

Baker D, Kjellström T, Calderon R, Pastides H (1999) Environmental epidemiology: a textbook on study methods and public health applications. Preliminary edition. World Health Organization, Malta

Banning A, Benfer M (2017) Drinking water uranium and potential health effects in the German Federal State of Bavaria. Int J Environ Res Public Health 14(8):E927. https://doi.org/10.3390/ijerph14080927

Bulka CM, Jones RM, Turyk ME, Stayner LT, Argos M (2016) Arsenic in drinking water and prostate cancer in Illinois counties: an ecologic study. Environ Res 148:450–456. https://doi.org/10.1016/j.envres.2016.04.030

Caranci N, Biggeri A, Grisotto L, Pacelli B, Spadea T, Costa G (2010) The Italian deprivation index at census block level: definition, description and association with general mortality. Epidemiol Prev 34(4):167–176

Cech I, Holguin AH, Littell AS, Henry JP, O'Connell J (1987) Health significance of chlorination byproducts in drinking water: the Houston experience. Int J Epidemiol 16(2):198–207. https://doi.org/10.1093/ije/16.2.198

Chiang CT, Lian IB, Su CC, Tsai KY, Lin YP, Chang TK (2010) Spatiotemporal trends in oral cancer mortality and potential risks associated with heavy metal content in Taiwan soil. Int J Environ Res Public Health 7(11):3916–3928. https://doi.org/10.3390/ijerph7113916

Christian WJ, Huang B, Rinehart J, Hopenhayn C (2011) Exploring geographic variation in lung cancer incidence in Kentucky using a spatial scan statistic: elevated risk in the Appalachian coal-mining region. Public Health Rep 126(6):789–796. https://doi.org/10.1177/003335491112600604

Colak EH, Yomralioglu T, Nisanci R, Yildirim V, Duran C (2015) Geostatistical analysis of the relationship between heavy metals in drinking water and cancer incidence in residential areas in the Black Sea region of Turkey. J Environ Health 77(6):86–93

Collman GW, Loomis DP, Sandler DP (1988) Radon-222 concentration in groundwater and cancer mortality in North Carolina. Int Arch Occup Environ Health 61(1–2):13–18

Crump KS, Ng TH, Cuddihy RG (1987) Cancer incidence patterns in the Denver metropolitan area in relation to the rocky flats plant. Am J Epidemiol 126(1):127–135. https://doi.org/10.1093/oxfordjournals.aje.a114644

Cui C, Wang B, Ren H, Wang Z (2019) Spatiotemporal variations in gastric cancer mortality and their relations to influencing factors in S County, China. Int J Environ Res Public Health 16(5): E784. https://doi.org/10.3390/ijerph16050784

Dahl C, Søgaard AJ, Tell GS, Flaten TP, Krogh T, Aamodt G, NOREPOS Core Research Group (2013) Is the quality of drinking water a risk factor for self-reported forearm fractures? Cohort of Norway. Osteoporos Int 24(2):541–551. https://doi.org/10.1007/s00198-012-1989-7

Dai D, Oyana TJ (2008) Spatial variations in the incidence of breast cancer and potential risks associated with soil dioxin contamination in Midland, Saginaw, and Bay Counties, Michigan, USA. Environ Health 7:49. https://doi.org/10.1186/1476-069X-7-49

Deener KCK, Sacks JD, Kirrane EF, Glenn BS, Gwinn MR, Bateson TF, Burke TA (2018) Epidemiology: a foundation of environmental decision making. J Expo Sci Environ Epidemiol 28(6):515–521. https://doi.org/10.1038/s41370-018-0059-4

Dolk H, Mertens B, Kleinschmidt I, Walls P, Shaddick G, Elliott P (1995) A standardisation approach to the control of socioeconomic confounding in small area studies of environment and health. J Epidemiol Community Health 49(Suppl 2):S9–S14. https://doi.org/10.1136/jech.49.suppl_2.s9

Dreiher J, Novack V, Barachana M, Yerushalmi R, Lugassy G, Shpilberg O (2005) Non-Hodgkin's lymphoma and residential proximity to toxic industrial waste in southern Israel. Haematologica 90(12):1709–1710

Edwards M, Johnson L, Mauer C, Barber R, Hall J, O'Bryant S (2014) Regional specific groundwater arsenic levels and neuropsychological functioning: a cross-sectional study. Int J Environ Health Res 24(6):546–557. https://doi.org/10.1080/09603123.2014.883591

European Commission (2016) Call for proposals for a pilot project on primary prevention courses for girls living in areas with higher risk of breast cancer. http://ec.europa.eu/research/participants/data/ref/other_eu_prog/other/hp/call-fiche/hp-call-fiche-pp2-5_en.pdf. Accessed 5 Aug 2019

Fei X, Lou Z, Christakos G, Ren Z, Liu Q, Lv X (2018) The association between heavy metal soil pollution and stomach cancer: a case study in Hangzhou City, China. Environ Geochem Health 40(6):2481–2490. https://doi.org/10.1007/s10653-018-0113-0

Getis A (2010) The analysis of spatial association by use of distance statistics. Geogr Anal 24 (3):189–206. https://doi.org/10.1111/j.1538-4632.1992.tb00261

Grilc E, Gale I, Veršič A, Žagar T, Sočan M (2015) Drinking water quality and the geospatial distribution of notified gastro-intestinal infections. Zdr Varst 54(3):194–203. https://doi.org/10.1515/sjph-2015-0028

Guajardo OA, Oyana TJ (2009) A critical assessment of geographic clusters of breast and lung cancer incidences among residents living near the Tittabawassee and Saginaw Rivers, Michigan, USA. J Environ Public Health 2009:316249. https://doi.org/10.1155/2009/316249

Guillaume E, Pornet C, Dejardin O, Launay L, Lillini R, Vercelli M, Marí-Dell'Olmo M, Fernández Fontelo A, Borrell C, Ribeiro AI, Pina MF, Mayer A, Delpierre C, Rachet B, Launoy G (2016) Development of a cross-cultural deprivation index in five European countries. J Epidemiol Community Health 70(5):493–499. https://doi.org/10.1136/jech-2015-205729

Hanchette C, Zhang CH, Schwartz GG (2018) Ovarian cancer incidence in the U.S. and toxic emissions from pulp and paper plants: a geospatial analysis. Int J Environ Res Public Health 15 (8). https://doi.org/10.3390/ijerph15081619

Hänninen O, Knol AB, Jantunen M, Lim TA, Conrad A, Rappolder M, Carrer P, Fanetti AC, Kim R, Buekers J, Torfs R, Iavarone I, Classen T, Hornberg C, Mekel OC, EBoDE Working

Group (2014) Environmental burden of disease in Europe: assessing nine risk factors in six countries. Environ Health Perspect 122(5):439–446. https://doi.org/10.1289/ehp.1206154

Hastie T, Tibshirani R (1990) Generalized additive models. Chapman and Hall, New York

Hendryx M, Conley J, Fedorko E, Luo J, Armistead M (2012) Permitted water pollution discharges and population cancer and non-cancer mortality: toxicity weights and upstream discharge effects in US rural-urban areas. Int J Health Geogr 11:9. https://doi.org/10.1186/1476-072X-11-9

Huang HH, Huang JY, Lung CC, Wu CL, Ho CC, Sun YH, Ko PC, Su SY, Chen SC, Liaw YP (2013) Cell-type specificity of lung cancer associated with low-dose soil heavy metal contamination in Taiwan: an ecological study. BMC Public Health 13:330. https://doi.org/10.1186/1471-2458-13-330

Jian Y, Messer LC, Jagai JS, Rappazzo KM, Gray CL, Grabich SC, Lobdell DT (2017) Associations between environmental quality and mortality in the contiguous United States, 2000-2005. Environ Health Perspect 125(3):355–362. https://doi.org/10.1289/EHP119

Knegt DE, Coughenour MB, Skidmore AK, Heitkönig IMA, Knox NM, Slotow R, Prins HHT (2010) Spatial autocorrelation and the scaling of species–environment relationships. Ecology 91(8):2455–2465. https://doi.org/10.1890/09-1359.1

Lillini R, Vercelli M (2019) The local socio-economic health deprivation index: methods and results. J Prev med Hyg 59(4 Suppl 2):E3–E10. https://doi.org/10.15167/2421-4248/jpmh2018.59.4s2.1170

Lin WC, Lin YP, Wang YC, Chang TK, Chiang LC (2014) Assessing and mapping spatial associations among oral cancer mortality rates, concentrations of heavy metals in soil, and land use types based on multiple scale data. Int J Environ Res Public Health 11(2):2148–2168. https://doi.org/10.3390/ijerph110202148

López-Abente G, Locutura-Rupérez J, Fernández-Navarro P, Martín-Méndez I, Bel-Lan A, Núñez O (2018a) Compositional analysis of topsoil metals and its associations with cancer mortality using spatial misaligned data. Environ Geochem Health 40(1):283–294. https://doi.org/10.1007/s10653-016-9904-3

López-Abente G, Núñez O, Fernández-Navarro P, Barros-Dios JM, Martín-Méndez I, Bel-Lan A, Locutura J, Quindós L, Sainz C, Ruano-Ravina A (2018b) Residential radon and cancer mortality in Galicia, Spain. Sci Total Environ 610-611:1125–1132. https://doi.org/10.1016/j.scitotenv.2017.08.144

McDermott S, Bao W, Aelion CM, Cai B, Lawson AB (2014) Does the metal content in soil around a pregnant woman's home increase the risk of low birth weight for her infant? Environ Geochem Health 36(6):1191–1197. https://doi.org/10.1007/s10653-014-9617-4

Messier KP, Serre ML (2017) Lung and stomach cancer associations with groundwater radon in North Carolina, USA. Int J Epidemiol 46(2):676–685. https://doi.org/10.1093/ije/dyw128

Monrad M, Ersbøll AK, Sørensen M, Baastrup R, Hansen B, Gammelmark A, Tjønneland A, Overvad K, Raaschou-Nielsen O (2017) Low-level arsenic in drinking water and risk of incident myocardial infarction: a cohort study. Environ Res 154:318–324. https://doi.org/10.1016/j.envres.2017.01.028

National Center for Biotechnology Information. U.S. National Library of Medicine. https://www.ncbi.nlm.nih.gov/pubmed. Accessed 5 Aug 2019

Nieder AM, MacKinnon JA, Fleming LE, Kearney G, Hu JJ, Sherman RL, Huang Y, Lee DJ (2009) Bladder cancer clusters in Florida: identifying populations at risk. J Urol 182(1):46–50. https://doi.org/10.1016/j.juro.2009.02.149

Núñez O, Fernández-Navarro P, Martín-Méndez I, Bel-Lan A, Locutura JF, López-Abente G (2016) Arsenic and chromium topsoil levels and cancer mortality in Spain. Environ Sci Pollut Res Int 23(17):17664–17675. https://doi.org/10.1007/s11356-016-6806-y

Núñez O, Fernández-Navarro P, Martín-Méndez I, Bel-Lan A, Locutura Rupérez JF, López-Abente G (2017) Association between heavy metal and metalloid levels in topsoil and cancer mortality in Spain. Environ Sci Pollut Res Int 24(8):7413–7421. https://doi.org/10.1007/s11356-017-8418-6

Oliver MA, Webster R (1990) Kriging: a method of interpolation for geographical information systems. Int J Geograph Inf Syst 4(3):313–332. https://doi.org/10.1080/02693799008941549

Pannullo F, Lee D, Waclawski E, Leyland AH (2016) How robust are the estimated effects of air pollution on health? Accounting for model uncertainty using Bayesian model averaging. Spat Spatiotemporal Epidemiol 18:53–62. https://doi.org/10.1016/j.sste.2016.04.001

Ren H, Wan X, Yang F, Shi X, Xu J, Zhuang D, Yang G (2014) Association between changing mortality of digestive tract cancers and water pollution: a case study in the Huai River Basin, China. Int J Environ Res Public Health 12(1):214–226. https://doi.org/10.3390/ijerph120100214

Richmond RE, Rickabaugh J, Huffman J, Epperly N (1987) Colorectal cancer mortality and incidence in Campbell County, Kentucky. South Med J 80(8):953–957. https://doi.org/10.1097/00007611-198708000-00005

Rodrigues EC, Assunção R (2012) Bayesian spatial models with a mixture neighborhood structure. J Multivar Anal 109:88–102. https://doi.org/10.1016/j.jmva.2012.02.017

Roh T, Lynch CF, Weyer P, Wang K, Kelly KM, Ludewig G (2017) Low-level arsenic exposure from drinking water is associated with prostate cancer in Iowa. Environ Res 159:338–343. https://doi.org/10.1016/j.envres.2017.08.026

Rue H, Martino S, Chopin N (2009) Approximate Bayesian inference for latent Gaussian models by using integrated nested Laplace approximations. J R Stat Soc Ser B Stat Methodol 71:319–392. https://doi.org/10.1111/j.1467-9868.2008.00700.x

Saint-Jacques N, Brown P, Nauta L, Boxall J, Parker L, Dummer TJB (2018) Estimating the risk of bladder and kidney cancer from exposure to low-levels of arsenic in drinking water, Nova Scotia, Canada. Environ Int 110:95–104. https://doi.org/10.1016/j.envint.2017.10.014

Samaniego FJ (2010) A comparison of the Bayesian and frequentist approaches to estimation. Springer, New York. https://doi.org/10.1007/978-1-4419-5941-6

Sánchez-Díaz G, Escobar F, Badland H, Arias-Merino G, Posada de la Paz M, Alonso-Ferreira V (2018) Geographic analysis of motor neuron disease mortality and heavy metals released to Rivers in Spain. Int J Environ Res Public Health 15(11):E2522. https://doi.org/10.3390/ijerph15112522

Selvin S, Shaw G, Schulman J, Merrill DW (1987) Spatial distribution of disease: three case studies. J Natl Cancer Inst 79(3):417–423

Stewart BW, Wild CP (2014) World Cancer report 2014. International Agency for Research on Cancer, Lyon

Su CC, Lin YY, Chang TK, Chiang CT, Chung JA, Hsu YY, Lian IB (2010) Incidence of oral cancer in relation to nickel and arsenic concentrations in farm soils of patients' residential areas in Taiwan. BMC Public Health 10:67. https://doi.org/10.1186/1471-2458-10-67

Thorpe N, Shirmohammadi A (2005) Herbicides and nitrates in groundwater of Maryland and childhood cancers: a geographic information systems approach. J Environ Sci Health C Environ Carcinog Ecotoxicol Rev 23(2):261–278. https://doi.org/10.1080/10590500500235001

Tutz G (2011) Poisson regression. In: Lovric M (ed) International encyclopedia of statistical science. Springer, Berlin. https://doi.org/10.1007/978-3-642-04898-2_450

Van Leeuwen JA, Waltner-Toews D, Abernathy T, Smit B, Shoukri M (1999) Associations between stomach cancer incidence and drinking water contamination with atrazine and nitrate in Ontario (Canada) agroecosystems, 1987-1991. Int J Epidemiol 28(5):836–840. https://doi.org/10.1093/ije/28.5.836

Wasaby Website. http://www.wasabysite.it/. Accessed 5 Aug 2019

Woods LM, Rachet B, Coleman MP (2005) Choice of geographic unit influences socioeconomic inequalities in breast cancer survival. Br J Cancer 92(7):1279–1282. https://doi.org/10.1038/sj.bjc.6602506

World Health Organization Website. Health policy page. http://www.euro.who.int/en/health-topics/health-policy. Accessed 5 Aug 2019

Printed in the United States
by Baker & Taylor Publisher Services